WATER, EARTH, AND FIRE

JONATHAN BERGER AND
JOHN W. SINTON

WATER, EARTH, AND FIRE

*Land Use and Environmental Planning
in the New Jersey Pine Barrens*

THE JOHNS HOPKINS UNIVERSITY PRESS, Baltimore and London

This book is dedicated to the memory of Rodney Koster, from Herman City, and of Narendra Juneja, whose maps of air, land, water, life, and location inspired the title and whose sense of logic and beauty is sadly missed.

The paper in this book is acid-free and meets the guidelines for permanence and durability of the Committee on Production Guidelines for Book Longevity of the Council on Library Resources.

The Johns Hopkins University Press, 701 West 40th Street, Baltimore, Maryland 21211
The Johns Hopkins Press Ltd, London

Photographic Credits. American Folklife Center, Library of Congress: p. 43. James F. Gandy, Jr., pp. 30, 46, 62, 69, 70, 104, 110, 124, 134. Hap Haven, pp. 81, 87, 201. Keystone Aerial Service, 74, 93. Arthur Rothstein, Farm Service Administration (Library of Congress), 7, 55, 77, 113. United States Department of Agriculture, 38. Title page art from Thomas Eakins, *Pushing for Rail*. The Metropolitan Museum of Art, Arthur H. Hearn Fund, 1916.

Library of Congress Cataloging in Publication Data
Berger, Jonathan.
    Water, earth, and fire.

    Bibliography: p.
    Includes index.
    1. Land use—New Jersey—Pine Barrens.   2. Natural resources—New Jersey—Pine Barrens.   3. Land use—New Jersey—Pine Barrens—Planning.   4. Environmental policy—New Jersey—Pine Barrens.   I. Sinton, John W.   II. Title.
    HD266.N52P563   1984   333.75′09749   84–47963
    ISBN 0–8018–2398–6 (alk. paper)

Contents

Illustrations

Acknowledgments

The people of the Pine Barrens have been extraordinarily generous in talking with us, and it is their material that forms the heart of this book. Of the several hundred residents we talked with, we would like to give special thanks to the following: Don Kleiner, Ted Von Bosse, Dan O'Conner, Albert Reeves, Bill, Clara, Noreen, and Bill Wills Jr., Budd Wilson, George Mick, Mary-Ann Thompson, Bill Haines Jr., Toby Green, Duke Galletta, Russ Clark, Marie Wynn, Will Grunow, Sam and Caren De Cou, Jack Cervetto, Bill Smith, John Neal, Joe King, Sam Hunt, Janice Sherwood, Gladys Eayre, Ed Hazelton, Leo and Hazel Landy, Danny Franchetti, and Mo Siegel.

Many of our colleagues gave valuable advice and have been kind enough to support our work in many ways, from editorial changes to financial support. We are particularly grateful to the Stockton State College Research and Professional Development Committee and to the David Berger Foundation for funding help and to the following for their advice: Dr. Silvan Tomkins, Dr. John Bennett, Dr. Daniel Bates, Prof. Ian McHarg, Dr. Daniel Rose, Dr. Arthur Johnson, Dr. William Gilmore, Dr. Joseph Rubenstein, Dr. David Fairbrothers, Dr. Angus Gillespie, Prof. Stephen Dunn, Dr. Nora Rubinstein, Dr. Max Silverstein, Dr. Elizabeth Marsh, Dr. Richard Smardon, Dr. David Kinsey, Mr. Philip Marrucci, Mr. Ben Coe, Mr. Hal Williams, Mr. Richard Regensburg, Mr. J. B. Jackson, Prof. Michael Laurie, Mr. Budd Wilson, Dr. Silas

Little, Ms. Sharon Coady, and Mr. George Pierson. In addition, Terry Moore and John Stokes, of the Pinelands Commission staff, and consultants John Rogers and Charles Siemen gave us valuable insights into the Pinelands planning process. We thank the Pinelands Commission for providing data for most of the maps and to the staff at Batsto for the use of the map of Wharton's water supply plan. Thanks also to Michael Worobec and Hap Haven, who designed and produced most of the maps, and to Hap Haven for his photographs; to mapmaker Patricia Coleman; to artist Anthony Hillman; and to Janice Feldstein, who cleaned up the manuscript and helped make it more readable.

Our editor, Wendy Harris, has been of inestimable help. Her insights and prodding have kept us at our task with more joy than despair, and her work has given the book more focus, organization, and directness than we had thought possible. We are also grateful to Cynthia Hotvedt, who designed the book, and Carol Ehrlich, who saw the book through its final stages.

Finally, for their moral support and their patient caring, we thank Juanita Bradshaw Sagan and our wives, Kit Wallace and Wendy Sinton.

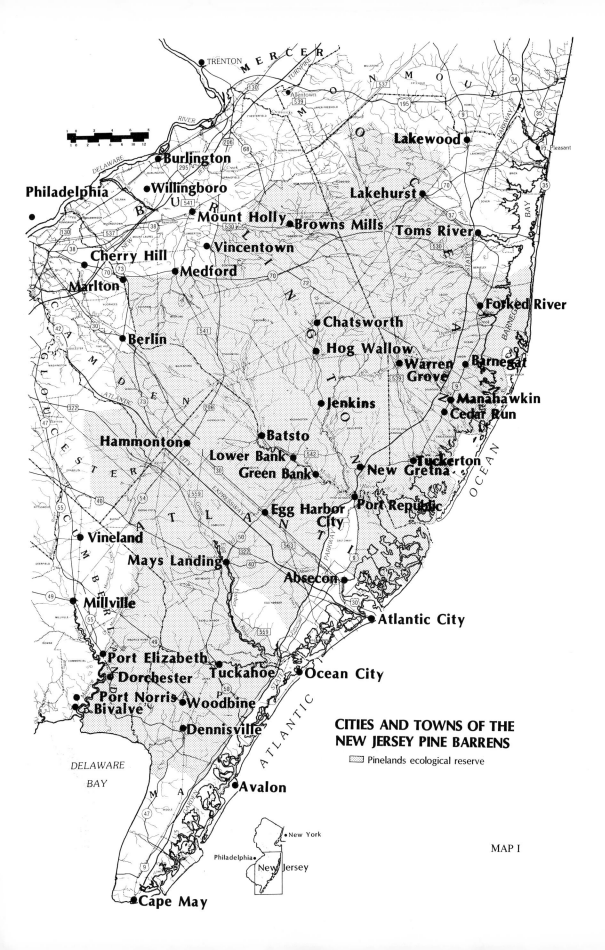

**CITIES AND TOWNS OF THE
NEW JERSEY PINE BARRENS**

Pinelands ecological reserve

MAP I

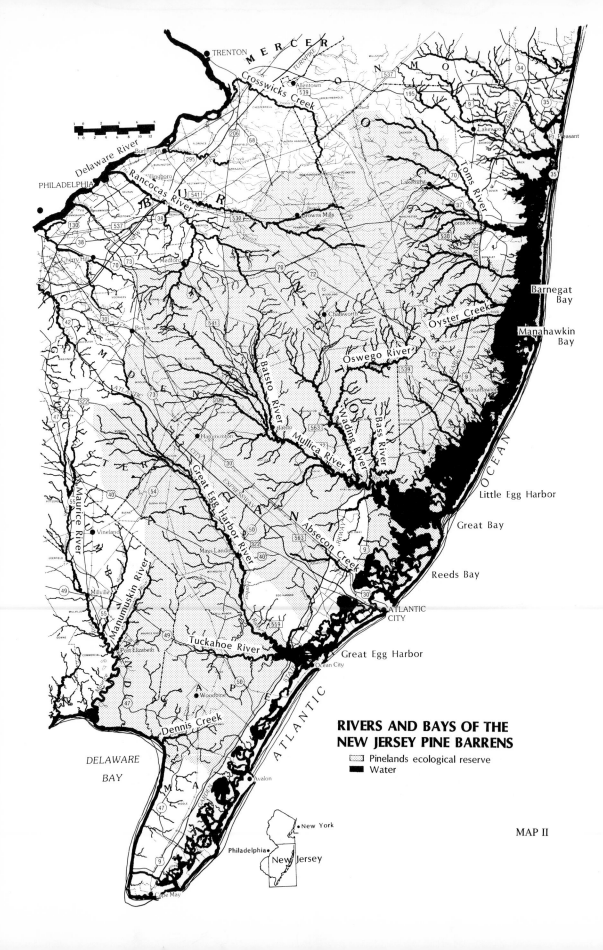

RIVERS AND BAYS OF THE
NEW JERSEY PINE BARRENS

Pinelands ecological reserve
Water

MAP II

Introduction

   We have written this book for both the planning community and interested general readers as an attempt to help develop more fertile ground for land-use planning. In our work we found that planners generally do not give sufficient attention to human ecology when developing plans. This book uses the Pine Barrens as an example of how planning can continue to move toward a more realistic understanding of a region, thereby creating a more useful plan that responds not only to place, but to people as well.

   Each major chapter of *Water, Earth, and Fire* begins with an introductory overview of the historical and ecological context of land-use patterns and the resources on which people depend. Readers then meet a series of individuals, people who represent several of the hundreds whom we interviewed during the time we worked as consultants for the Pinelands Commission, the planning agency for the Pine Barrens. In other words, we first present the view of scientists and then the view of residents who represent the dominant cultural themes of the place, as the subheadings of each chapter suggest: ''The Collective Memory and Seasonal Cycle,'' ''The Heart and Conscience of the Community,'' ''Seclusion and the Special Place,'' ''Insiders and Outsiders,'' and so forth. Together the observations from both the scientific (operational) and insider (cognized) models form a new way of looking at the realities of a place; this is the heart of the socionatural system.

The first chapter sets the Pine Barrens in the larger context of American regional planning, because what is happening in the Pine Barrens will affect future regional planning in America. We introduce the Pine Barrens through a short history of their changing land-use patterns and community and social structures. This chapter serves as a backdrop so readers can better understand the following three chapters, which carefully explain the links between people, their resources, and their use of the land.

Chapter 2, on water, explores the intertwining of people and resource use from saltwater to freshwater environments and the complexity and richness of the relationships. Chapter 3, on earth, defines relationships between farmers and soil resources. The fourth chapter, fire, deals with fire ecology and forest resources. The forest lands symbolize the uniqueness of the Pines, although they are in fact but one of its three major elements.

The fifth chapter takes material from the first four and translates it into propositions for developing a regional plan for the Pine Barrens. Chapter 6, the conclusion, summarizes our findings. We make no pretense in this book to completeness; we cannot cover all aspects of planning that become part of a master plan. Such a task would require a second volume to update and analyze all consultants' reports on archaeology, economics, hydrology, law, zoology, and other topics. We hope with this work to advance the efforts of environmental planning toward a better synthesis of material, a more cogent way to understand a region and to translate that understanding into a planning context.

Since the material we present in *Water, Earth, and Fire* is a planning aid, not a panacea, the principles we derive in chapter 5 constitute an expansion of our abilities to synthesize and reveal information rather than a full-fledged plan. To collect the kind of data we present is not expensive compared to the costs of gathering hydrologic data, but it does take patience and training. While the inclusion of socionatural concepts (the combination of social and environmental factors) into planning does not make the resolution of conflicts any simpler, the synthesis will depict the conflicts more clearly and, most important, will make for more effective plans that respond to an enlarged understanding of the realities of a place. Plans, to be effective, must reflect the interactions between people and their environments, and this is much easier in a ten-square-mile township than in a thousand-square mile region. Yet the principle is the same. It is more complex and time-consuming to treat a region with care and precision, but it is worth it.

Last, the information in the next four chapters can help dispel a planner's arrogance that only he or she really understands a region and how to treat it. The beauty and complexity of the Pine Barrens should awe the officials in charge of the region's future. The deep understanding of any place can leave no room in a planner's mind for cynicism.

WATER, EARTH, AND FIRE

Chapter 1

THE PINE BARRENS AND REGIONAL PLANNING

The Physical Setting

The Pine Barrens of New Jersey are special. Dwarf trees in the Pine Plains, huge amounts of fresh water underground, rare plants and animals, and indigenous landscapes have sprung from three hundred years of interaction between people and place. The Pine Barrens are important for their very location—a million sparsely settled acres on the borders of America's most densely populated corridors (maps 1a–d). Their unique nature is the subject of much of this book, but the implications of what we know and what planners have done in the past four years in the Pine Barrens reach beyond the region itself.

The Pinelands is the official designation of what most residents still call the Pine Barrens. The region comprises more than 1,500 square miles between New York and Philadelphia, small in comparison to many areas of national concern, but more than 20 percent of the entire state of New Jersey and important to national and international planning communities. As the letterhead of the Pinelands Commission, the region's planning entity, proudly proclaims, the region is our nation's first national reserve, a region of mixed urban, suburban, agricultural, forest, and wildland uses; the purpose of the national reserve is to encourage all levels of government to devise innovative ways to plan rationally for the region. Official international recognition came in 1983, when the region became part of the United Nations' Man and the Biosphere Program, which

PHYSIOGRAPHIC
REGIONS OF EASTERN
AND CENTRAL USA

Pine Barrens

PLAINS
Coastal
Great
Central lowlands
St. Lawrence lowland
PLATEAUS
Piedmont
Appalachian Interior
Ozark
New England
MOUNTAINS
Blue Ridge
Adirondack
Ouachita

© 1974.

MAP Ia. From *Natural Regions of the United States and Canada*, by Charles Hunt.
W. H. Freeman and Company.

focuses international attention on sensitive areas of the earth's biosphere. There are 225 biosphere reserves in 62 countries, 40 of them in the United States; the Pine Barrens are one of two in the Atlantic Coastal Plain from Long Island to Florida, the other being South Carolina's Congaree Swamp.

Despite such recognition, a plane trip across the Pine Barrens cannot impress the casual visitor. The one million acres of the Pines seem flat, undistinctive, uninteresting. Auto trips do not reveal much more. Driving along the Garden State Parkway or state route 72, passersby from North Jersey or the Delaware Valley may notice little except the mileposts that tell them their distance from the beaches of the Jersey Shore. Ask people who have traveled these routes where the Plains or Pygmy Pines are, and most cannot tell you. The Pine Barrens reveal their diversity and secrets only to the careful, to people who live there or those observant enough to look with awareness.

The scrubby trees all appear similar to untrained eyes, but people who know the Pines can easily tell where water lies by the presence of white cedar or maple. Oaks, on the other hand, mean the soil is dry. There are probably more oaks than pines in the Pine Barrens and more than a dozen oak species and hybrids. Pitch pines are scattered among the oaks and in pure stands. Pines make up for their lack of species diversity by diversifying their forms from straight and tall to crooked and broad; in the Plains they are grotesquely twisted, fit for bonsai specimens.

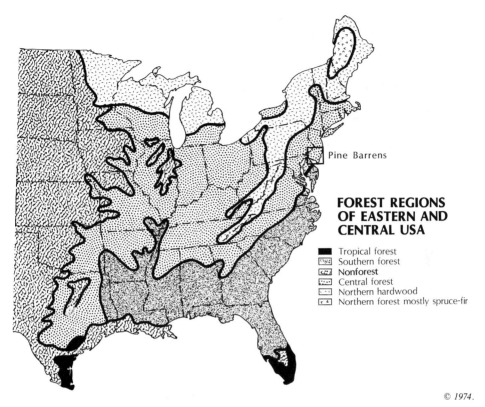

Pine Barrens

**FOREST REGIONS
OF EASTERN AND
CENTRAL USA**

■ Tropical forest
▨ Southern forest
▨ Nonforest
▨ Central forest
▨ Northern hardwood
▨ Northern forest mostly spruce-fir

© *1974.*

MAP 1*b*. (After American Forest Institute). From *Natural Regions of the United States and Canada,* by Charles Hunt. W. H. Freeman and Company.

One must often think small in the Pine Barrens. Erosion and deposition, gigantic processes in the Rockies and midwestern states, are reduced to miniature in the Barrens, where river gorges are ten feet deep and alluvial plains are dead sandy deltas of the Delaware River. Even deer, which weigh 115 pounds and more in the neighboring Delaware Valley, may weigh only 80 pounds in these forests.

The Pine Barrens seem vast because of the accretion of many small things: a million acres of forests with small trees; more than 17 trillion gallons of water in just one aquifer made from raindrops that filter through the soil; extraordinary numbers of rare and endangered plants and animals, none larger than an eight-foot snake, most smaller than a green frog or shorter than a lead pencil.

The colors of the Pines appear dull to the unobservant, but they are subtly beautiful: somber cedar, bright pine, light and dark oak greens, tones one can pick out from a black-and-white aerial photograph. Autumn brings a full spectrum of browns and extraordinary splashes of maroon from black gum, scarlets from the red maple, and orange from sassafras. Differences in climate are much less subtle. While one might desiccate in the sandy heat of the Plains in August, one might almost suffocate in the fetid humidity of a cedar swamp the same day.

People born in the Pines know all this because Pine Barrens landscapes do not exist apart from their residents. Cedar stands that look ancient have actually been cut five

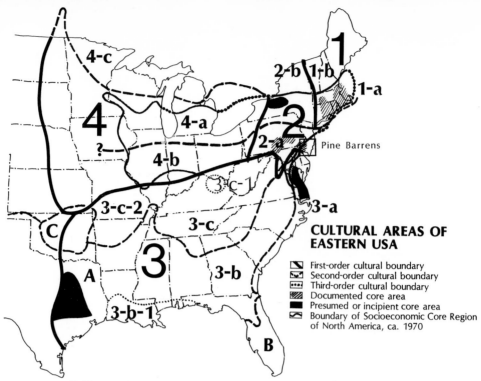

CULTURAL AREAS OF EASTERN USA

Legend
- First-order cultural boundary
- Second-order cultural boundary
- Third-order cultural boundary
- Documented core area
- Presumed or incipient core area
- Boundary of Socioeconomic Core Region of North America, ca. 1970

MAP 1c. Wilbur Zelinsky, *The Cultural Geography of the United States*, © 1973, p. 118. Reprinted by permission of Prentice-Hall, Inc., Englewood Cliffs, N.J.

Region	Settlement and Formation*	Major Sources of Culture†
New England		
1-a Nuclear New England	1620–1750	England
1-b Northern New England	1750–1830	Nuclear New England, England
The Midland		
2-a Pennsylvania Region	1682–1850	England and Wales, Rhineland, Ulster, 19th-century Europe
2-b New York Region, or New England Extended	1624–1830	Great Britain, New England, 19th-century Europe, The Netherlands
The South		
3-a Early British Colonial South	1607–1750	England, Africa, British West Indies
3-b Lowland or Deep South	1700–1850	Great Britain, Africa, Midland, Early British Colonial South, aborigines
3-b-1 French Louisiana	1700–1760	France, Deep South, Africa, French West Indies
3-c Upland South	1700–1850	Midland, Lowland South, Great Britain
3-c-1 The Bluegrass	1700–1800	Upland South, Lowland South
3-c-2 The Ozarks	1820–1860	Upland South, Lowland South, Lower Middle West
The Middle West		
4-a Upper Middle West	1800–1880	New England Extended, New England, 19th-century Europe, British Canada
4-b Lower Middle West	1790–1870	Midland, Upland South, New England Extended, 19th-century Europe
4-c Cutover Area	1850–1900	Upper Middle West, 19th-century Europe

*approximate dates
†listed in order of importance

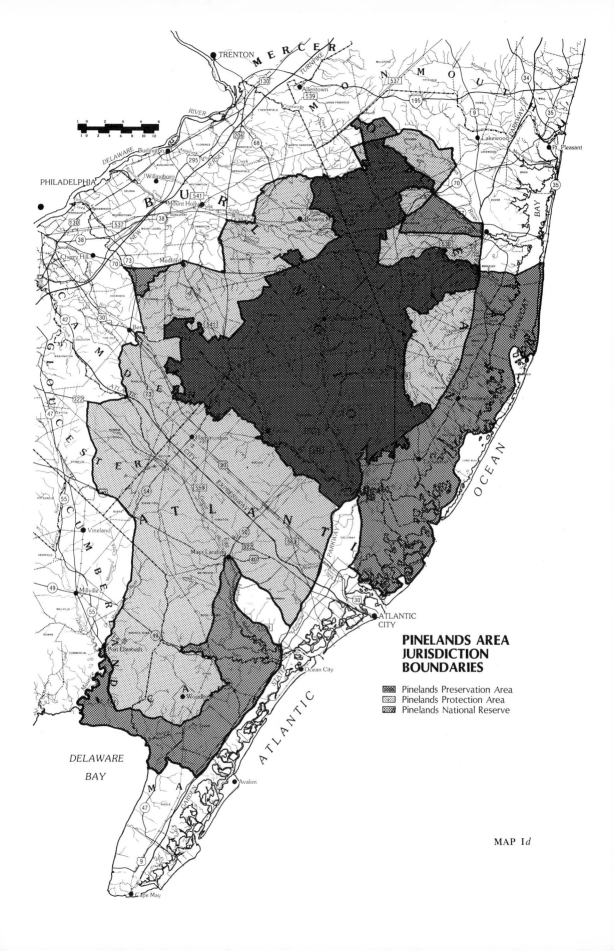

**PINELANDS AREA
JURISDICTION
BOUNDARIES**

▨ Pinelands Preservation Area
░ Pinelands Protection Area
▨ Pinelands National Reserve

MAP I*d*

times in the past 200 years; pine forests were burned or harvested every 15 to 20 years. These are not wildlands, but landscapes that humans have husbanded and exploited these past 2,000 years. The observant eye will pick out fire-cracked rock from Indian camps, white sands where charcoal pits existed, poison ivy (called just ''poison'' in the Pines) that suggests the foundations of a house, and ripples on the soil that belie former blueberry fields. The old memory will be reminded of gardens and farms now grown up in forest, grave sites hidden by oak leaves where a pet horse or dog lies buried, and places where grandfathers cut wood or built boats time out of mind. The people of the Pines are intimate with their families, their communities, and their landscapes. The Pines are kind to people who wish to lead their own lives and not take from the land and water more than these resources can give.

A Short Land-Use History of the Pine Barrens

One can discover from Pine Barrens landscape history three major patterns of settlement. The first settlers came down the coast and up the rivers in the seventeenth, eighteenth, and nineteenth centuries. At first Dutch, English, and New Englanders, later northern Europeans and blacks, they based their economic activities on wood, agriculture, fishing, hunting, gathering, and seafaring. The second pattern of settlement involved the railroads, which cut through the heart of the Pines in the 1850s and 1860s. With railroads came immigrant groups—Germans, Italians, and Eastern European Jews—who established towns that replicated those in their homelands in what must have seemed to them like isolated spots in the Pines, such as Woodbine, Egg Harbor City, and Vineland. And with railroads came links from the new agricultural areas to the cities, enabling truck farming, orcharding, and berrying to grow as commercial enterprises. The railroads also brought tourists, not only to Atlantic City, but also to places like Chatsworth and other communities in the middle of the Pines, where speculators sold quarter-acre lots to unsuspecting metropolites, who, on seeing the lots, threw up their hands and left only paper streets amid the encroaching forestland (photo 1). The third settlement pattern appeared after World War II and can now be seen scattered across the Pine Barrens as suburban and second-home developments— foreign bodies placed on an indigenous landscape.

The river towns of Tuckahoe and Corbin City nicely illustrate the first set of patterns. Charles Hartman of Millville, a surveyor and engineer, mapped in detail the eighteenth- and nineteenth-century patterns of this area in the southeast corner of the Pine Barrens, fifteen miles southwest of Atlantic City. The most outstanding features on the map are the nine sawmills in a fifty-square-mile area, all built between 1737 and 1812. The original sawmills and ponds, around which small communities of three or four households gathered, served the larger coastal and estuarine villages with their shipbuilding and commercial activities. Tuckahoe was typical of such villages: an 1834 gazeteer (Gordon 1834, 254) described it thus:

> Tuckahoe, port town on both sides of the Tuckahoe river, over which there is a bridge, 10 miles above the sea, 46 miles S.E. from Woodbury [the county seat], and by post-route 192 from Washington [D.C.]; contains some 20 buildings, 3 taverns, several stores. It is a place of considerable trade in wood, lumber, and ship building. The land immediately on the river is good, but a short distance from it, is swampy and low.

PHOTO I. *A Failed Real Estate Development*

Between 1760 and 1880 the forests provided wood not only for ship and house building but also for iron furnaces, forges, and glass houses. The iron-furnace and forge towns lasted in general from thirty to ninety years, and while they did not display the permanence of the river towns, they were more substantial in their heyday. Aetna furnace, five miles upriver from Tuckahoe, had died by the time Gordon wrote his 1834 gazeteer, but nearby was its sister, Weymouth furnace, still under blast when Gordon was writing. He described it as follows (1834, 263):

> Weymouth, blast furnace, forge and village, in Hamilton township, Gloucester Co., upon the Great Egg Harbour river, about 5 miles above the head of navigation. The furnace makes about 900 tons of castings anually: the forge having four fires and two hammers, makes about 200 tons bar iron, immediately from the ore. There are also a grist and a saw mill, and buildings for the workmen, of whom 100 are constantly employed about the works, and the persons depending upon them for subsistence, average 600 annually. There are 85,000 acres of land pertaining to this establishment. . . . The works have a superabundant supply of water, during all seasons of the year.

During the two hundred years between the first settlement and the advent of railroads, the forests were cut, cleared, and fired several times, and speculators succeeded and failed over and over again with rural industries and woodlots. This pattern held throughout the Pine Barrens; an 1850 resident of Browns Mills in the northwest would have recognized immediately the similarity of land-use patterns at Tuckahoe.

Archetypal landownership patterns and vernacular architecture from 1700 to 1850 still exist in the Pines. The coastal towns, at least the southern ones that have not been suburbanized, still have the same street patterns and old houses, including what Elizabeth Marsh (1982, 185–87) called the South Jersey house.

> The distinguishing feature of the South Jersey house is that the front door is in the middle and that there is only one window on each side of the door. The building may be one, one-and-a-half, or two stories high. The ridgepole is always parallel to the road and the chimney, if it remains, is inside the house at the end. This design is found elsewhere, but it persisted longer in the Pinelands and along the coast. That is why I call the type ''South Jersey.''
>
> Houses tell you something about their builders. What does the South Jersey house say to us? This house is small and modest, yet symmetrical, almost elegant in its proportions. It was built and rebuilt without change for about 150 years. This house reflects a people who were self-respecting, in spite of their meagre resources, with pride in their workmanship. It tells us of the people of the Pinelands and coastal villages, caught in their old ways while the rest of the nation was going west.

Another relic from the first stage of settlement is the extensive amount of unbroken forest lands from iron-furnace days when furnaces, with their 80,000-acre properties, backed one on another. At one point in the first half of the nineteenth century, a Philadelphia Quaker family, the Richards, owned hundreds of thousands of acres from central Burlington County to southern Atlantic County, all forestlands that fed their furnaces. It was, therefore, comparatively easy for Joseph Wharton in the 1870s, a generation after the death of the iron furnaces, to buy large chunks of undeveloped land in that same area and for Wharton's heirs in 1950 to sell that same land in one piece to the state of New Jersey. In addition to the large-holder patterns were a multitude of parcels belonging to small scam developers and speculators, some wholly un-scrupulous. As a result, title to undeveloped land is often unclear in this region. The large- and small-holder patterns lay side by side throughout the Pines and strongly influenced current development patterns.

Immediately after the Civil War, railway companies completed the lines they had begun in the 1850s. The railroads, along with the rest of the Industrial Revolution, changed patterns in most of the Pine Barrens. The most lasting impact was to cut the Pines into two sections north and south of the Mullica River. Because the two major east-west lines ran to the tourist resorts of Atlantic City and Cape May, a rail net developed throughout the southern section, while the northern section had no east-west link, only one that ran north-south through the central, least populated areas. Tuckahoe, a good example of the southern region, had one north-south railroad, and an east-west line just to the south of it.

The rail net south of the Mullica spawned truck farms and orchards, health spas and boondoggle real-estate developments, small clothing industries and ethnic settlements. Although large tracts of forestland remained for charcoal makers and sphagnum-moss gatherers to work, much of it was turned into fields—the higher ground for peaches, apples, tomatoes, cucumbers, and sweet and white potatoes, and the wetter ground for blueberries, when they were cultivated after World War I. In addition, cranberry fever hit the region in the 1860s; landowners and lessees turned old millponds and swamps into cranberry bogs. An enormous amount of cedar was no doubt lost as the trees were drowned and cut to make way for new bogs.

The railroad also created new speculative and settlement opportunities, as it did throughout America. Italians, Germans, and Eastern European Jews settled in ethnic communities in the southern region. Some came as railroad workers, while others bought or were given land by German Protestant and Jewish philanthopists. The Jews of Woodbine, the Germans of Egg Harbor City, and the Italians of Hammonton thrived and built self-contained agricultural and industrial communities; their economic eminence held until after World War II. Speculators, including railway companies, bought cheap forestland and advertised small lots to anyone foolish enough to buy them. Speculative developments littered the maps of the southern Pine Barrens, almost all of them failures like Montefiore, Gigantic City, and Little Italy. In the late nineteenth century, New York and Philadelphia newspapers commonly gave away small pieces of ground in the Pine Barrens with new subscriptions.

North of the Mullica River, speculators and cranberry operators also flourished. As happened in the southern Pines, few developments succeeded; Paisley and Fruitland, for example, are now mere curiosities. But the cranberries did succeed, and, because the watersheds were somewhat larger and competing land uses fewer, extensive cranberry bogs still exist. The old rural industrial towns and outlying villages, however, are now lost in the woods. No railroads came to reinvigorate the economy, so the northern Pine Barrens slowly returned to woods, swamps, and bogs after the Civil War. Woodcutters, colliers, gatherers, hunters, and berry cultivators exploited the section of the Pines no one else wanted.

After World War II two major processes occurred: the small industrial complexes of the southern section began to die as industry continued to concentrate in metropolitan areas, and, with the advent of the population boom and of high-speed roads and automobiles, suburban, second-home, and retirement developments all became feasible.

The northern fringes of the Pine Barrens have borne the brunt of suburbanization since 1945, not so much because land was cheaper, but because the area was closer to Philadelphia and New York. Although the north-central section, known as the "core," still displays some landscapes left from rural industrial and railroad periods, suburbs and retirement communities have nibbled off the borders all along the coastal and Delaware Valley sections; at times suburbs have leapfrogged over existing developments into the heartland, as has the retirement community of Leisure Village.

The southern section of the Pines, being farther from any metropolis, witnessed little suburbanization until recently, and has seen its small industrial base die. Woodbine lost all its industries, and Egg Harbor is but a shadow of its old self; only Hammonton retains a solid but diminished industrial base. Along the coastal strip are miles of vacation homes, but few such developments have appeared inland. The southern section still bears the heavy imprint of the second wave of settlement from the railroad, but, because of its proximity to Atlantic City casinos, it is finally under suburbanization pressure.

Cultural Point and Counterpoint

Despite the diversity of communities in the Pine Barrens, the area forms, with some suburban exceptions, a cultural region with its own ethos, its own cultural characteristics. Kai Erikson's (1976) valuable study of Buffalo Creek, West Virginia, elucidates something of the cultural traditions of people in the Pine Barrens.

> To speak of culture is to speak of elements that help shape human behavior—the inhibitions that govern it from the inside, the rules that control it from the outside, the languages and philosophies that serve to edit a people's experience of life, the customs and rituals that help define how one person should relate to another. To speak of culture is to speak of those forces that promote uniformity of thought and action. But there are other forces at work in culture too. . . . The identifying motifs of a culture are not just the *core values* to which people pay homage but also the *lines of point and counterpoint* along which they diverge. That is, the term "culture" refers not only to the customary ways in which a people induce conformity in behavior and outlook but the customary ways in which they organize diversity. In this view, every human culture can be visualized, if only in part, as a kind of theater in which certain contrary tendencies are played out.
>
> At the societal level, these contrasting tendencies are experienced as a form of *differentiation*. The people of any culture sort themselves into a wide variety of groups and segments, each of them sharing something of the larger culture at the same time that each tries to fashion modes of living peculiar to itself. When these differences develop into competition for power or goods or are based on antagonisms of long standing, the result is apt to be social conflict; but when these differences develop into an implicit agreement to apportion the work of the culture as well as its rewards on the basis of the contrasting qualities that each group represents, the result is apt to be a form of complementarity. (81–83)

A good deal of complementarity still exists in the Pine Barrens. The existence of such harmony in work and land-use patterns jars contemporary skeptical minds, but this does not make the harmonious patterns any less real. There are three primary reasons for the continued existence of harmony. First, until recently, most of the Pines were immune to the kinds of population pressure and landscape changes that transformed adjacent areas (we have seen why in the previous section describing the region's land-use history). Pressures for rapid change destroy social and land-use fabrics because the old-timers cannot adjust and the newcomers cannot integrate. Second, unlike places like Appalachia, the resources of the Pine Barrens are renewable. People depend on a wide variety of resources and markets, and the extraction of resources, save sand and gravel mining, does not damage the landscape. Last, the Pine Barrens, with few exceptions, are not isolated, the result being an economically and culturally integrated relationship of urban and rural patterns.

To illustrate the relationship between Pineys and the world, a woman with deep Piney roots told us a story about the catastrophic docking of the zeppelin, the Hindenburg, at hangar #1 in Lakehurst in 1938, when it developed a hydrogen leak and blew up. "Anti-Nazi feelings were running high," said our friend, "expecially about the way they treated Jews." So her uncle and a small group of friends took their rifles and waited in a swamp near hangar #1 for the Hindenburg to dock. As it glided toward the docking station, the men aimed and shot at the swastika on the tail of the zeppelin. They missed the swastika, and the next thing they knew, the Hindenburg went up in flames.

Subcultures in the Pines are no more isolated, nor are they startlingly different, from rural cultures in other mid-Atlantic sections. If such rural-urban relationships do not foster uniqueness in the Pines, they do provide people with the ability to respond to social change in more flexible ways than people isolated in the Appalachians.

A cultural tradition given much credence in other studies on the Pine Barrens is the myth of the isolated Pineys in their wilderness. Joan Goldstein (1981, 39) gives the classic definition as

> the remote Pineys who were reported to have been descended from runaway Hessian soldiers during the American Revolution; or runaway slaves from the South, who, finding the forest of New Jersey, believed they had reached free Canada; or independent frontiersmen, privateers, and Indians.

The most pernicious root of the myth lies in the study of the Kallikak family conducted in the first decade of this century at the Vineland Training School by H. H. Goddard and his research assistant, Elizabeth Kite. Stephen Jay Gould (1981, 168) studied Goddard's work and found that

> Goddard discovered a stock of paupers and ne'er-do-wells in the pine barrens of New Jersey and traced their ancestry back to the illicit union of an upstanding man with a supposedly feeble-minded tavern wench. The same man later married a worthy Quakeress and started another line composed wholly of upstanding citizens. Since the progenitor had fathered both a good and a bad line, Goddard combined the Greek words for beauty (*kallos*) and bad (*kakos*), and awarded him the pseudonym Martin Kallikak.

In an absurd attempt to support the budding eugenics movement, Goddard (1912) and Kite traced the *kakos* family history through nine generations of alcoholism, blindness, criminality, deafness, epilepsy, insanity, syphilis, sexual immorality, tuberculosis, and feeblemindedness. His book concludes:

> The Kallikak family presents a natural experiment in heredity. From this we conclude that feeble-mindedness is largely responsible for these social sores. Feeble-mindedness is hereditary and transmitted as surely as any other character. . . . In considering the question of care, segregation through colonization seems in the present state of our knowledge, to be the ideal method. Sterilization may be accepted as makeshift, as a help to solve this problem because the conditions [of continued sexual liaisons] have become so intolerable. (116–17)

In fact, the Goddard study was a hoax. Goddard's photographs, which he used as supporting evidence, ''were phonied by inserting heavy dark lines to give eyes and mouths their diabolical appearance'' (Gould 1981, 171). Even the tongues hanging out of the depraved mouths had been forged. Still, the stigma of the *kakos* line attached itself to all residents of the Pine Barrens.

Like all cultural groups, Pineys have folktales, and the tales have enhanced the myth of Pineys as a remote and isolated group. In the thirties and forties, Herbert Halpert (1947) wrote a landmark doctoral dissertation on Piney folklore and unintentionally contributed to the mystique. The tall stories about the Jersey Devil and hyperbolic anecdotes about the mysterious woods showed more about Pineys' love of folktales than their ignorance of the outside world.

Halpert, however, was but one of many visitors who came to listen to stories. The Pineys added to their own reknown by embellishing, ornamenting, and inventing new tales. As Janice Sherwood said to the author (J.S.), "Why tell 'em the truth when they'll believe a lie?" Another Piney friend once said,

> Boy, you should'a heard the stories we'd tell old Dr. Beck [the late Reverend Henry Charlton Beck who was an author and reteller of tales]. He was an awful nice man, but, Lord, he'd believe anything. We'd set it up so's I'd tell him one story up here in Green Bank and old H⸺ would tell him a totally different one down Lower Bank.

The myth of isolation and wilderness, then, was fed from both within and without. Little wonder that McPhee's (1968) Fred Brown became the archetypal Piney or that Goldstein (1981, 47–48) could write that

> the life style of the Pineys, living as they did in harmony with nature, as hunters and gatherers . . . helped keep them removed from the political and social institutions of the mass society. . . . They are, from their own description [that of Gladys Eayre] frightened and timid.

In fact Fred Brown lived much of his life in a cabin a few hundred yards from a major road on the edge of New Jersey's largest and most modern cranberry enterprise, and all but one of his children live as solid middle-class citizens, integral parts of their communities. Gladys Eayre, despite her gruff exterior, is a well read and politically savvy women who lives more on the bay than in the woods.

What, then, is a Piney? Nora Rubinstein (1983, 191) discovered during three years of fieldwork that

> "Pineyness" was based on geographical location at various stages in life, with birth-place being of greatest significance, [then] ancestry, age, occupation, economic status, family ties, and an amorphous quality many comprehend, but few can determine. It is an attitude, a way of being in the world, an essence or quality not included in the legislative description. . . . [My] search for the elusive Piney has come full circle, from geography and language, ancestry and occupation to an affective sense—a feeling for family, but most important, for the land, for the experience of 'being' in the Pines. It is just over the horizon, or as Janice Sherwood said, "a little deeper in the woods than you are."

It is important to understand that while "being in the Pines" is essential to being a Piney, the Pine Barrens have never truly been isolated geographically or culturally.

Individuals and families who live outside the mainstream, therefore, do so more from choice than force.

Having disposed of some myths about Pineys, we can now discuss a series of cultural values common to people in the Pine Barrens, the most important of which are: topophilia, or love of place; the central role of the family in social transactions; Christian morality (primarily Methodism); participation in seasonal activities such as farming, hunting, fishing, and trapping; and high participation levels in local politics and in such voluntary organizations as the farm bureau, civic organizations, and particularly ambulance and fire companies.

Topophilia is the attachment to physical places—cemeteries, cedar forests, rivers and lakes, bays, and the old home place. While there is a good deal of mobility in and out of the Pines, those who leave for jobs in the city or a dream farm in Maine return often. It is true that people return to, or even decide to stay in, the Pine Barrens for family reasons, but love of the physical setting in which they grew up plays a major role. There are many residents of the Pines who grew up in some other place but after falling in love with the Pine Barrens came to live in them.

Family ties are strong in the region. Family structure tends to be overtly patriarchal in that men have jobs, hold public office, and speak for the household in the presence of outsiders, while women run the households and raise the children. Individual psychological problems are solved, if at all, within the family or in conjunction with the local pastor. Most churches in the Pines are Methodist, some are Presbyterian or, in Italian and Hispanic areas, Catholic. The church supports the traditional morality of strong family ties, charity (or neighborliness), honest labor, and temperance. Alcoholism, while common, appears no more or less a problem than in neighboring urban or suburban sections, and is frowned on; that is, people are expected to carry on their work and family life in a responsible manner, not succumb to liquor.

Pine Barrens residents, both men and woman, place a high value on seasonal activities that reinforces their love of place. Good farmers or gardeners are held in as high esteem as good hunters. Men tend to participate more often in such activities as hunting or fishing, but both sexes often take walks together, pick roadside flowers, and tend gardens together.

Participation in political and voluntary organizations on the local level is high. The most important of these are churches for women and volunteer fire and ambulance companies or gunning clubs for men. Local farm bureaus and granges are active in farming areas, and the lake communities have their own series of organizations, while Rotary and Lions clubs are common in the suburbs. Pine Barrens residents tend to be very knowledgeable about local politics even though they may attend public meetings infrequently unless something important to the community is at stake.

Lines of point and counterpoint are particularly important in establishing and holding the social and ecological balance of the Pine Barrens, as is true in almost any region. Here the characteristic point and counterpoint consists of: resource sharing versus competition for resources, protection of common resources versus individual property rights, love of tradition versus eagerness for change, and urban versus rural life styles.

Despite the implications of the term barrens, the Pine Barrens are rich in resources, from pinecones and laurel for wreaths, to hunting lands, to farmland, to clams, oysters, and fish. How people share these resources is part of the subject of this book, but the whole question of sharing is one of tension. Surely, the most important shared resource is water, and individuals cannot use water unless they make formal or informal agreements to share it. Similarly, clammers have worked out systems whereby they can individually get enough clams to make a living while still leaving enough for others. Unfortunately, the balance in this particular subsystem is being rapidly destroyed by increasing population pressure along the coast; newcomers have never been part of the traditional sharing-competition systems. We shall see a part of this drama unfold in the next chapter.

While topophilia is a central value in the Pines, so, too, is the tenet that one can do with one's property as one sees fit—a person has the right to sell his or her land for the highest price even though the land may have been used for decades as a gathering place by the community. When few people were interested in land or resources in the region, this point and counterpoint lay hidden, but now it crops up often in the Pines. On the one hand, a person may be angry that the regional power of the Pinelands Commission will not allow him to sell his land to a developer for top dollar; on the other, he may be just as angry that his neighbor would want to do the same, disfiguring the neighborhood. On an individual level, topophilia makes people act ambivalently, some claim hypocritically; on the social level, it has split towns in half, some people promoting development, others preservation.

Such arguments about development and preservation can be seen in the light of two other contrasting tendencies. One is love of tradition versus a search for change. There are those who cling dearly to traditional rural styles that include open space and low density development, while others, even in the same family, see tradition as a road to poverty and lack of progress. In stylistic terms one finds this controversy played out as urban versus rural values, whereby urban or, more often, suburban landscapes signify progress while rural ones suggest backwardness. Unlike people in Appalachia, residents of the Pine Barrens have always had, at least on an individual level, a choice between the country and the city because Philadelphia and New York, even in the eighteenth century, were never more than two days' travel. That the choice has always been available suggests that such tension has always existed. Until recently, however, there was not enough demand inside the region to suburbanize it, so individuals who preferred urban life had to move to those places. That is no longer true.

The tension along these last axes is almost taut at present in some sections. Where people find a new balance will determine the extent to which these subregions of the Pine Barrens can survive as areas with an indigenous culture and whether the landscapes can remain intact as change slowly continues. The culture and landscape are of a piece, and the destruction of one dooms the other.

We now move to a different level of description—the townships in which people live. Although people's activities transcend municipal boundaries, the town still forms an essential part of their lives.

Pinelands Townships

In New Jersey the township, or municipality, is a crucial political entity, second in importance only to the state; it has traditionally exercised local planning and zoning powers. Every township in the Pine Barrens has its own personality. Some towns in the

Pines are larger than 90 square miles, and their character is revealed in the smaller communities or villages within it. Other townships, smaller than 10 square miles, are cohesive political and social units. Historic land-use patterns and contemporary resource use are determining factors, and geographical location plays a major role in a town's makeup. Proximity to the coastal and Delaware Valley borders determines the amount of development pressure any town bears (maps 2a, 2b). In an analysis of those pressures, Alan Mallach (1980b, 33) found that

> the most significant conclusion is that economic activity, of the sort linked to present or potential development, is limited in the extreme within the Pinelands area. Furthermore, it would appear that a substantial part of the residential development in the Pinelands is linked to economic activity taking place outside the boundaries of the Pinelands region.

The individuals whom readers meet in this book were chosen to represent the diverse socionatural systems of the Pines, and the townships from which they come are as different as the individuals themselves. Table 1 gives socioeconomic data from the 1980 census on towns in which our main protagonists live. It is clear that ethnic composition, job descriptions, economic status, and real-estate values differ significantly from town to town. For example, the neighboring towns of Medford and Southampton have similar ethnic bases but are quite different in terms of jobs and economics.

Since our next chapter is on water, we start with the coastal townships, first in Atlantic County. Don Zehner lives in Galloway, a municipality of more than ninety square miles, originally settled along its coast in the communities of Leeds Point, Oceanville, Smithville, and Conovertown. Egg Harbor City Germans settled its western area in the mid-nineteenth century. Because of its proximity to Atlantic City and a sewer main constructed along the main coastal road, Galloway is undergoing intensive suburbanization of its coastal sections, where 75 percent of the population lives. It is a predominantly white, middle-class, German-English-Irish town with a significant Italian minority. As suburbanization continues, the percentage of white-collar workers is increasing while the blue-collar percentage remains stationary, and farming in the western sections continues to decline. Sixty-eight percent of the residents were born in New Jersey, higher than the state average, but the level of education is close to the New Jersey average. The sections of the township in the state Pinelands area are changing only slowly, partly as a result of Pinelands Commission regulation but mostly in the face of recession and slow demand for housing and other uses in that area. There is no "Galloway" in Galloway; it remains a collection of old coastal villages and nineteenth- and twentieth-century communities, most of which have their own schools, volunteer fire companies, and voluntary associations.

Port Republic, where Ted Von Bosse, the crabber, lives, was one of Galloway's coastal communities until the first decade of this century, when it became an eight-square-mile "city." It is a cohesive, white community that looks like a New England town on a tidal creek over which looms the tall steeple of the Methodist church. "Port," as residents call it, was founded near the site of a small Revolutionary War naval battle at Chestnut Neck and was a fishing and shipbuilding center. Because its aesthetics and cohesiveness are rare in South Jersey, its land is in high demand by upper-middle-class and rich families coming into the area as a result of casino

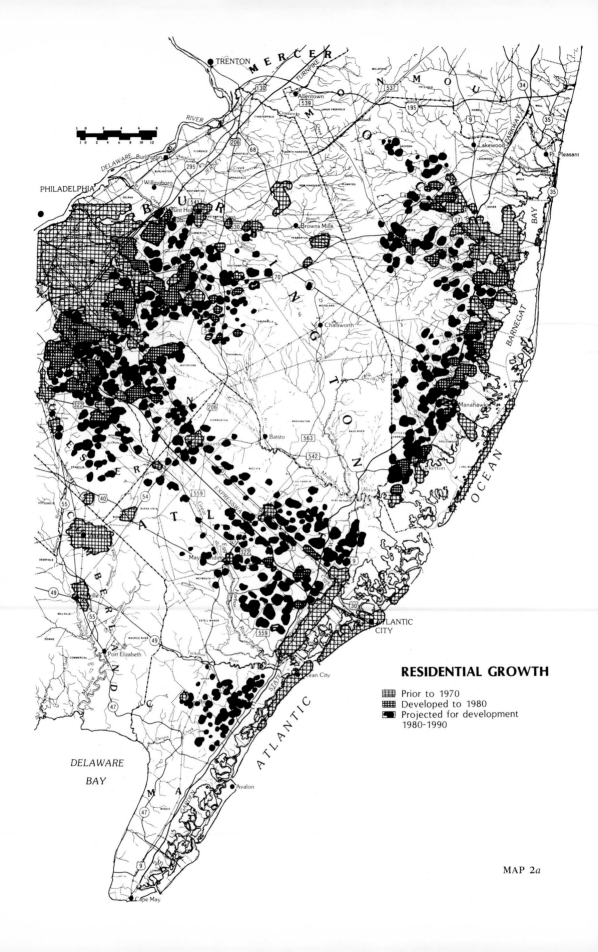

RESIDENTIAL GROWTH

Prior to 1970
Developed to 1980
Projected for development
1980-1990

MAP 2a

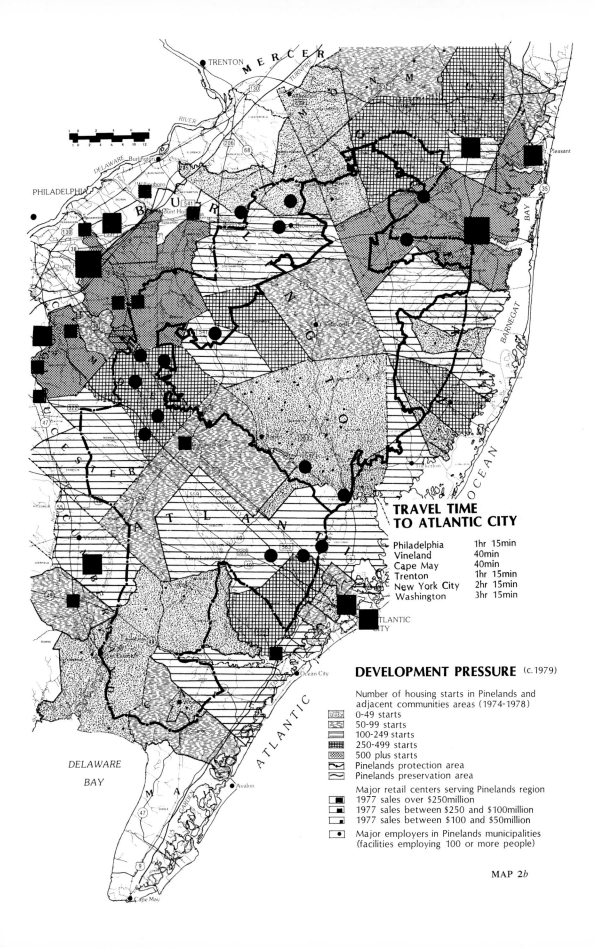

TRAVEL TIME TO ATLANTIC CITY

Philadelphia	1hr 15min
Vineland	40min
Cape May	40min
Trenton	1hr 15min
New York City	2hr 15min
Washington	3hr 15min

DEVELOPMENT PRESSURE (c.1979)

Number of housing starts in Pinelands and adjacent communities areas (1974-1978)

0-49 starts
50-99 starts
100-249 starts
250-499 starts
500 plus starts
Pinelands protection area
Pinelands preservation area

Major retail centers serving Pinelands region

1977 sales over $250million
1977 sales between $250 and $100million
1977 sales between $100 and $50million

Major employers in Pinelands municipalities
(facilities employing 100 or more people)

MAP 2*b*

TABLE I. *1980 Socioeconomic Census Data*

Total Population	White %	Black %	Hispanic %	Mean # children per married couple	English and mixed lr. ethnic %	German and mixed Ger. ethnic %	Irish and mixed lr. ethnic %	Italian and mixed lt. ethnic %	% Born in N.J.	% Unemployed	% over 16 not in labor force	Mgmt., exec., professional %	Tech., sales, administrative	Service	Farming, forestry, fishing	Precision pro. crafts	Blue Collar	Over 18 no high school diploma	High School Diploma completed	Some post-2nd School	Post-college study	Median family income	Median Non-rental unit	Median gross rnt.
New Jersey 7,364,823	79	13	7	1.9	13	19	19	19	56	7	37	26	33	12	0.8 (24,982)ª	12	17	31	37	27	5	22,907	67,985	270
Egg Harbor City 4,618 11 sq. miles	73	11	15	2.0	11	31	17	19	62	13	40	15	29	11	2 (37)	14	29	44	34	19	2	19,911	44,383	211
Galloway 12,100 92 sq. miles	92	5	2	2.0	22	33	25	14	68	7	34	18	29	18	2 (138)	16	16	31	40	24	5	20,074	55,598	359
Hammonton 12,298 42 sq. miles	89	2	8	2.0	10	13	12	55	68	11	37	20	27	12	4 (204)	14	23	41	40	15	4	19,554	51,840	253
Maurice River 4,577 95 sq. miles	84	14	2	2.0	33	22	18	7	75	14	54	13	20	15	2 (31)	18	32	54	35	10	1	17,038	25,563	249
Medford 17,622 40 sq. miles	99	—	—	2.0	29	33	28	17	50	4	34	38	34	9	0.3 (59)	11	7	15	35	22	11	29,878	84,639	330
Mullica 5,243 57 sq. miles	79	8	13	2.2	21	27	21	18	56	8	42	19	23	15	3 (59)	21	19	39	39	17	4	20,725	45,954	286
Port Republic 913 8 sq. miles	100	—	—	2.1	36	33	20	8	64	5	35	25	23	22	4 (18)	15	10	26	45	22	7	24,167	67,909	308
Southampton 8,808 43 sq. miles	99	—	1	1.9	32	34	23	10	53	6	55	25	27	14	1 (40)	14	19	32	43	19	6	19,042	52,219	263
Stafford 10,385 47 sq. miles	99	—	—	2.0	20	27	23	18	65	11	45	21	27	17	2 (63)	20	12	35	43	18	3	17,190	45,200	250
Washington 830 107 sq. miles	88	—	11	2.4	28	35	27	5	73	5	46	19	18	18	12 (38)	14	19	41	38	17	3	17,026	44,352	275

SOURCE: *Department of Commerce, Bureau of the Census (1980).*
ªTotal number.

gambling. Port is now a mixture of old coastal families dating back to the eighteenth century, college professors, and casino workers.

Twenty miles up the coast from Port Republic is Stafford Township, named after the abandoned village of Stafford Forge; nothing remains of the early nineteenth-century forge. The township, in which Jack Cervetto and Ed Hazelton live, is a complex collection of Pineys, baymen, developers, seasonal residents on the barrier island, and white-and blue-collar workers. Stafford contains examples of every development type found in the Pine Barrens: Manahawkin, the old coastal town, a lake development, a lagoon development, a military installation, some small farms, and extensive open forest with the old community of Warren Grove at its center. Voluntary organizations range from the Rotary Club to church groups to fire companies to gunning clubs to a group dedicated to the preservation of Warren Grove's landscapes. Vacation-home development threatened to transform lives and landscapes at a rapid pace in the 1960s, but recessions and coastal regulation slowed that pace, so one sees in the 1980 census indications that the rate of change has decreased; job types and educational and economic status suggest a mix of old-fashioned and suburban life-styles. For example, low median income and property values indicate that living costs are inexpensive, as is common in rural areas.

Some of the most rural areas of the Pine Barrens are in the extreme south, such as the coastal/river town of Maurice River. Albert Reeves, the waterman, lives across the Maurice River, in Commercial Township, whose socioeconomic profile parallels that of Maurice River Township. The township has had a depressed economy since the severe decline of the oyster and shipbuilding industries in the middle of this century. Unemployment remains high, especially among blacks whose roots are in Maryland's eastern shore and whose livelihoods depended almost exclusively on oysters. The median value of a house in Maurice River, for example, is half one in Stafford and more than 300 percent less than one in Medford.

The other rural community we discuss at length in this book is Washington Township, in the northern core of the Pines, where George Mick and Bill Haines live. There is no "Washington" in the township; it used to consist of a few houses and a tavern where post roads cross, but it died a hundred years ago, as did other hamlets like Quaker Bridge and The Mount. Washington Township had three times its present population in the mid-nineteenth century because it was the center of rural industrial villages like Batsto, Martha, and Harrisville, all of them gone except for Batsto, which is now an outdoor museum. Still, 12 percent of the work force, three times that of any other town, make their living from the water, earth, and forests of the region. Washington's population is tiny—only 830 people in more than 100 square miles, but the state owns three-quarters of the land in the town as Wharton State Forest. Most people come from old families in the area, and the significant Hispanic minority works the cranberry bogs; many of them have lived in the Pines for two generations. Change proceeds slowly in Washington, and demand for development, even in the river communities of Green Bank and Lower Bank, remains low.

Thirty miles northwest of Washington, straddling a major transportation artery, are the towns of Medford and Southampton, examples of classic suburban encroachment onto historical eastern farming landscapes. Mary-Ann Thompson and the de Cous live in these towns, which retain some of their dairy and cranberry landscapes, South-ampton more so than Medford. While Medford's population is significantly wealthier and better educated than that of any other Pine Barrens township, it still has small industries, namely, the Medford Mills and a cannery, although most of the workers live in adjacent communities. Politics are dominated by residents of the lake communities, but the Quaker families, who created the original farming and commercial landscapes, still hold some power. Southampton's politics, on the other hand, have recently been dominated by older residents who moved into the retirement community of Leisure Town. Southampton's agricultural fields, many of which are leased, are under severe development pressure, and the historic Quaker center of Vincentown no longer exerts the influence over the township it traditionally did.

In the northwest corner of Atlantic County, twenty miles due south of Medford, is Russ Clark's hometown, Hammonton, originally a small Yankee community, which was transformed by Italian immigrants in the mid-nineteenth century into a vital farming center with some small industries. It retains its vitality to this day; farmers own their land and rarely lease it, community cohesiveness is strong, and the town's ethos is decidedly middle class and conservative, much like a classic small midwestern town. Of all Pine Barrens townships caught between the pressures to develop and to preserve, Hammonton has the greatest chance to maintain its balance and to control the pace of change.

Not so fortunate is Egg Harbor City, settled by Germans at the same time Italians settled Hammonton. Marie Wynn lives there, and Willard Grunow farms near its borders in the section of Galloway that Egg Harbor Germans settled. Because Egg Harbor's economy depended more on small industry than farming, it was more susceptible to the economic decline that followed industrial displacement after World War II.

Egg Harbor's influence also extends west into Mullica Township, which separates Egg Harbor City and Hammonton. There is no place called Mullica except the river that forms the town's northern border, where the first major settlement of Pleasant Mills originated. Like Washington, its sister to the north, Mullica is essentially a town of rivers, farms, and forests, the place where Leo and Hazel Landy live. It is a township that ecologically and culturally links the northern to the southern forest sections of the Pine Barrens.

These are only 10 of the 52 municipalities in the Pinelands region. The diversity of the Pinelands surprises even those who live there. Our interest is not simply in the wonder of diversity, but in the ways by which people share and exploit their resources across town boundaries and in the balances they have struck between point and counterpoint. This assemblage of differences defines the Pine Barrens as a cultural as well as an ecological region.

The Political and Legislative Setting

Over the past twenty years the Pine Barrens have been in demand as never before, not because of demands for drinking water for urban areas (Joseph Wharton thought of that more than two generations ago), not because of rare and endangered species (naturalists have known these for a hundred years), not because of land speculation (that has been occurring these last two hundred years). It is because the Pines are the last remaining large open area for easily accessible recreation and development in the megalopolitan corridor between Boston and Richmond. For the first time, large-scale residential developments have sprung up in the heart of the Pines, and one can no longer walk or canoe without hearing the sound of off-road vehicles, nor can one any longer rediscover last year's Pine snake dens or Pine Barrens gentians, because collectors have shipped them to markets in New York or Miami. For these reasons, land-use management in the Pine Barrens is no longer a matter for local politics alone, or even state politics. Competing land-use demands in the Pines have finally made the area a national and international concern.

Rumblings of major demands for development were heard in the early 1960s when some developers dreamed of a gigantic jetport in the Pines. By then a Pinelands Regional Planning Board already existed, created in 1964. Shortly after the jetport was suggested, the National Park Service and Philadelphia Academy of Natural Sciences began an ecological study of the central Pine Barrens, which resulted in the late Jack McCormick's *The Pine Barrens: A Preliminary Ecological Inventory* (1970). By 1968 both Ocean and Burlington counties had become involved in the Pinelands planning process, the results of which led to the creation of the Pinelands Environmental Council in 1972.

As interest grew, two other federal efforts, one by the Bureau of Outdoor Recreation in 1975 and another by the National Park Service in 1978, helped lead to the 1978 National Park and Recreation Act, section 503 of which designated the Pine Barrens as America's first national reserve. Meanwhile, the Pinelands Review Committee was created by Governor Brendan Byrne as a state planning advisory agency. It was the forerunner of the present Pinelands Commission established in 1979 as a result of New Jersey's Pinelands Protection Act of 1979, legislation that paralleled the 1978 federal act.

With the birth of an officially designated Pinelands region (see map 1d), the Pinelands Commission was established as a regional planning agency with considerable power to create preservation and development zones. The boundaries of the region over which the commission presided were delineated partly in response to the federal legislation that had established the concept of the national reserve. Other criteria for boundaries were ecological, political, administrative, and urban. On the basis of its unique soils and vegetation, biologists had established the outlines of the Pine Barrens in the first two decades of this century (Harshberger 1916; Stone 1911); the work of Jack McCormick (1970, 1979) set contemporary ecological boundary lines. Certain political interests had to be served in the New Jersey legislature in order to pass the 1979 Pinelands Protection Act, so some boundaries were shifted to account for them. In addition, New Jersey had in 1971 established a coastal zone with its own mandate and bureaucracy (now called the Division of Coastal Resources), so much of the eastern and southern boundaries followed that line. Last, the urbanized and suburbanized sections of the Jersey shore, such as Atlantic City, were excluded from the Pinelands act (and, for that matter, from discussion in this book). While the contemporary Pine Barrens show the clear impact of these shore communities on planning, resources, and economics, they are primarily products of the postwar period. Prior to that, tourist communities were small seasonal markets for regional resources, but not sources of population pressure or cultural change. Only recently have sections of the coast experienced year-round commercial and residential use.

As with other regional planning agencies, the Pinelands Commission and its staff knew immediately that it had major problems to solve: interpretation of the intent of both federal and state legislation, establishment of the legal standing of its decisions, a short (one-year) time frame in which to complete a plan for the million-acre region, and the creation of a plan that would help resolve conflicts between competing uses and users of land. If the plan were to prove workable, the residents of the Pine Barrens would, in the long run, have to accept it.

The Pinelands Commission adopted a Comprehensive Management Plan for the million-acre region in 1980 and is presently in the process of enforcing that plan. Despite early general agreement about the preservation and development goals of the legislation on the part of most landowners, realtors, bankers, farmers, and local politicians, the planning effort remains controversial and widely criticized by professionals and citizens alike. Some areas are overly regulated, while others have become dumping grounds for future development. There is a sense that the plan does not fit the place.

The lessons we are learning from planning in the Pine Barrens form many of the conclusions of this book. In this chapter we look at how regional planners can view not only the Pine Barrens but also other special areas within the larger context of American planning. Although this book is not a text on regional planning, we do need to explain what we understand as the purpose of planning and how it has been practiced recently in the United States. Readers who wish to probe more deeply into the technical aspects of planning can refer to the Appendix.

Tasks of Regional Planning

A planner's tasks are to understand a place by revealing its human and environmental aspects, to establish clearly where conflicts and cooperation occur, and to find techniques to solve conflicts and help create as healthy an environment as possible for present and future residents. No better definition of the revelatory functions of planning has yet been written than that of Benton MacKaye (1928, 147–49), creator of the Appalachian Trail, in his book *The New Exploration,* written almost sixty years ago.

"To command nature," said Plato, "we must first obey her. . . ." Here we have the function of every sort of "planner"; it is primarily to uncover, reveal, and visualize—not alone his own ideas but nature's; not merely to formulate the desire of man, but to reveal the limits thereto imposed by a greater power. Thus, in fine, planning is two things: (1) an accurate formulation of our own desires—the specific knowledge of what it is we want; and (2) an accurate revelation of the limits, and the opportunities, imposed and bequeathed to us by nature. Planning is a scientific charting and picturing of the thing which man desires and which the eternal forces will permit. The basic achievement of planning is to make potentialities visible. But this is not enough. Visibility is only part of the revelation. The mold must be rendered not only visible but audible. It must be *heard* as well as *seen.* The regional planner, in revealing a given mold or environment, must portray not alone for the sense of sight, but for that of hearing also. Indeed he must portray in terms of all the senses. If he would portray for us a possible colonial village in New England, he must portray it in terms of the three dimensions—and of the five senses. He must in his imagination see the Common, and hear the church bell, and smell the lilac blossoms, and contact the green-shuttered houses: he must also see and hear and contact the human activity. The individual art of painting consists in revelation in the realm of eyesight; the individual art of music consists in revelation in the realm of hearing; but the synthetic art of developing environment consists in revelation in the realm of all the senses. Planning is revelation—an all-round revelation.

Over the last fifteen years planners have developed quite accurate ways to reveal "the limits, and the opportunities, imposed and bequeathed to us by nature." The process, as it is presently practiced, was first described in Ian McHarg's book *Design with Nature* (1969) and requires planners to investigate geological, hydrological, pedological, biological, and social information about a region. The data are then mapped, and the planner and his or her clients can see what environmental opportunities and constraints exist on the landscape—where the waterlogged sites are, or the habitats for rare plant and animal species, or the recharge areas where rainwater percolates back into the soil to recharge residents' well water. This process is what is generally referred to as environmental or ecological planning.

The more difficult problem for planners is to derive "an accurate formulation of our own desires—the specific knowledge of what it is we want." Generally, a regional planner finds this in the goals and objectives of legislation for a particular region; as has been evident in Pinelands planning, however, legislative goals for the region are not always compatible. One goal, for example, is to provide for appropriate development, while another calls for the preservation of land and resources. Just as critical is an understanding of exactly who "we" are and how to handle the often ambivalent nature of our needs, such as the wish to preserve private-property rights while at the same time preserving the character of a community.

Most planning difficulties occur in the attempts to analyze the human parts of systems; it is easier for scientists to describe the nonhuman environment than the people and cultures that exist in a region. In other words, most planners have trouble including humans in ecological planning. Sometimes planning attempts are limited to the projection of population growth in order to plan for highways, schools, hospitals, police, and other physical services called the infrastructure. To population projections are often added census data on the age, religious, ethnic, and economic profiles of a place, and finally information on its political structure—how decisions are made and who makes them. Rarely have planners been able to synthesize the natural and human components of a region. This book is an attempt to help us continue to grope toward just that synthesis. In his article titled "Human Ecological Planning" (1981, 110), McHarg expressed the problems of environmental planners succinctly:

Human ecological planning is a cumbersome and graceless title. Remedy however while possible is distant. When it becomes accepted that no ecosystem can be studied without reference to man then we may abandon the "human" descriptor and revert to "ecological planning." Better still when planning always considers interacting biophysical and cultural processes then we can dispense with the distinction of "ecological" and simply employ the word planning. However that state is far in the future as most planning today excludes the physical and biological sciences, ecology, ethnography, anthropology, epidemiology and concentrates upon economics and sociology.

Thus we are stuck at the first—the revelatory—task in planning. If we cannot understand the combined human and environmental system (what we call the socionatural system), then we will have difficulty clarifying the conflicts and congruences, and, through our ignorance, we will help create, rather than mitigate, such conflicts.

Earlier we mentioned MacKaye's concept of all-round revelation. A knowledge of the socionatural systems and of residents' understanding of the region is vital to planning's revelatory task. This understanding comes not only from a scientific overview of a region, but also from the voices of the residents themselves—how their lives fit or do not fit annual and generational cycles, what resources they do or do not use, the historic development of their landscapes, their sacred places. It is the insiders' view that planners most often neglect, because most do not yet have a framework into which they can incorporate such information, and insiders' views often conflict. A scientific model is convenient because it allows a planner to piece together sets of predigested data, whether biological, sociological, or political, and these data constitute one way to see the ''reality'' of a region. Such realities are malleable to planning techniques because the data can be mapped, political and economic interest groups can be polled to discover where power lies, and economic data can be used to support or reject various alternatives so that one can create a legal framework for the reality.

But there is more to the real world, and that is the view of insiders. All of us, of course, are insiders in our own communities, and all of us are from time to time ambivalent about the place in which we live. Each sees it differently depending on age or economic situation or even mood. However, each community, indeed each region, has a set of uses and values attached to it by its residents. So reality can also be seen as an assemblage of the views of insiders such as we portray in this book. The insiders' realities, on the other hand, are anathema to many planners because they stymie quick decisions and obscure the neatness of planning solutions; they muddle the scientific model. Our attempt in *Water, Earth, and Fire* is to show that we can understand a region like the Pine Barrens as a total larger than the sum of its parts, and, that while this may be a more complex model than the one most planners are accustomed to, it helps create more realistic and, therefore, more useful plans.

How so? First, the use of socionatural systems gives us a deeper understanding of the intricacies of a region, the ways in which people and their resources have adapted to each other over time, the differences among parts of the region, and the impacts of change on various sections. Thus we emphasize that a region is neither nature alone nor man alone, but a result of the interaction of the two. In chapter 5 we discuss the delineation of subregions in the Pine Barrens as an alternative to the Pinelands Comprehensive Management Plan's use of uniform zoning districts that encompass sections of the region very different from one another. A planner cannot conceive, much less delineate, subregions without reference to socionatural systems.

Likewise, the impacts of fast- or slow-paced change, which a planner must always gauge, cannot be interpreted without an understanding that goes beyond traditional science. The impacts of rapid development in one section of the Pine Barrens, near Atlantic City, for example, are not the same as those in another one close to Philadelphia. Some subregions in New York City's sphere of influence in Ocean County have provided good examples of the adverse impacts of rapid development; it, therefore, does not make good planning sense to force other subregions to undergo similar transformations, as the Pinelands Plan requires. Neither does it make good sense to prevent development in some subregions that have been accustomed for decades to slow, if not steady, growth, especially if that growth does not countermand the legislative goal of preservation. In short, the use of socionatural systems leads to the kind of detail that planners need to make more careful decisions and, thereby, respond to legislative goals more accurately and flexibly.

Once a planner gains a better all-round revelation, he or she has access to more fitting alternatives. For example, if the goals of Pinelands legislation are both to preserve and to develop the region, an understanding of the subregions provides opportunities for one or both of those goals to be administered appropriately for each subregion. As we shall see, farmland in Hammonton ought not to receive the same treatment as farmland in New Lisbon. The Pinelands plan requires each of the fifty-two municipalities to conform individually to standard zones and densities, so it might be argued that flexibility is built into the plan by the give-and-take conformance process. But if the zones and densities are not based on accurate information, if they fall far short of an all-round revelation, then the result will be faulty.

The use of socionatural systems and the view of insiders will not cure all of planning's problems. It cannot prevent people from getting angry or losing the value of land in which they invested. It cannot set boundaries, although it can help rationalize them; it cannot change political processes, although it might elucidate them; it cannot change, but it might stretch, the legal framework in which planning occurs. And the revelations of socionatural systems will certainly not dispense with conflict, but can give planners a more accurate picture of the substance of conflicts and where they will most likely occur. Some legislators will always complain that the intent of their legislation was subverted in their home state or region. Insiders and outsiders are bound to disagree simply because they stand on different ground. Some years ago, when one of the authors (J.S.) was camping in Maine, he said to a local resident, "This is beautiful country. I want it to stay just like this." To this the Mainer replied, "If that's what you want, why don't you just buy it?"

In a broader sense, the incorporation of the concept of socionatural systems can lead to better public involvement in a plan because residents will be involved from the beginning due to the very nature of the data collected. Planners will often be able to avoid such destructive decisions as that of the National Park Service to purchase a large section of Ohio's Cuyahoga Valley. In a June 1983 "Frontline" program, the reporter, Jessica Savitch, summed up the Cuyahoga problem:

> Now the people who created the park were idealists. They believed that a park "for all people, for all time" was for the greater good. They believed that a park should preserve a place of peace and tranquility for the many people who would come here from our over-crowded cities. But what about the local community, the people we've just met. Should the park be created at their expense? That is not what Congress originally intended. When the law was being written, Congressman Seiberling, the author of the legislation, and a man whose own home was in the park said . . . , "We're not just talking about the conservation of a piece of land, we're talking about the conservation of people as well. In planning the park, the people must be considered as a resource as well as the trees and flowers and birds and waterfalls." (Rubinstein 1983, 56–57)

In that ''Frontline'' program, Cleveland *Press* reporter Peter Almond poignantly mused, ''I wondered whether the National Park Service, and those in Washington, understood what it was they were going to create here in the Cuyahoga Valley.'' (Rubinstein 1983; 59) Lack of understanding and tunnel vision will more likely destroy than preserve landscapes.

If one understands a place more deeply and accurately, both from inside and outside, then one can see arguments about land uses and alternative futures in a more sophisticated manner, and the voices of clammers, woodcutters, and other ordinary citizens will be heard as clearly as those of preservationists and developers. Conflict resolution thus becomes a more democratic and, to the participants, acceptable practice; the definition of the ''public'' can more accurately reflect the idea of public involvement, and planners will no longer have to rely solely on the most outspoken groups or individuals, be they preservationists or developers.

Chapter 2

WATER

T he Pine Barrens are an undulating, sandy, gravelly cushion of water (maps 3, 4). Where a stream slowly meanders and where bogs and swamps occur, water is at the surface of the cushion. In the lowlands bordering streams and bogs, water is no deeper than a foot below the surface. Even in the driest uplands, surface water is never more than a mile away. We must understand three influential properties of Pine Barrens water to understand human use of the region: the water is acid and corrosive; it is interconnected; and it is generally clean and plentiful. These properties have provided opportunities for and constraints on human use and have caused many conflicts over control of the water resources.

The sand of the Barrens does not retard for long the flow of water into the ground. While nature left few minerals in the sand at the surface of the Pines, close to the surface and in deeper deposits it left the so-called green sands of glauconitic clays, deposited by an earlier ocean inundation. These sands are rich in iron, and when the corrosive groundwater passes through iron-rich sands, it picks up the iron and returns it to the surface in bogs and swamps (Crerar et al. 1978). Here, in the slow-moving backwaters, bacteria fix the iron and make it available for human use.

The water movement that brings iron to the surface also determines the nature and properties of the soil and links upland to tidal bay, surface to subsurface, and shallow

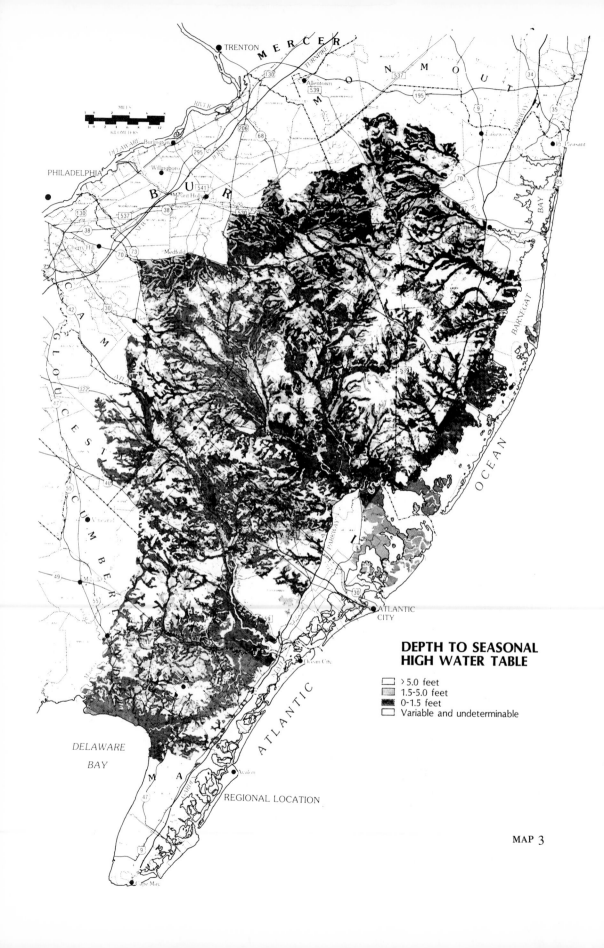

**DEPTH TO SEASONAL
HIGH WATER TABLE**

> 5.0 feet
1.5-5.0 feet
0-1.5 feet
Variable and undeterminable

REGIONAL LOCATION

MAP 3

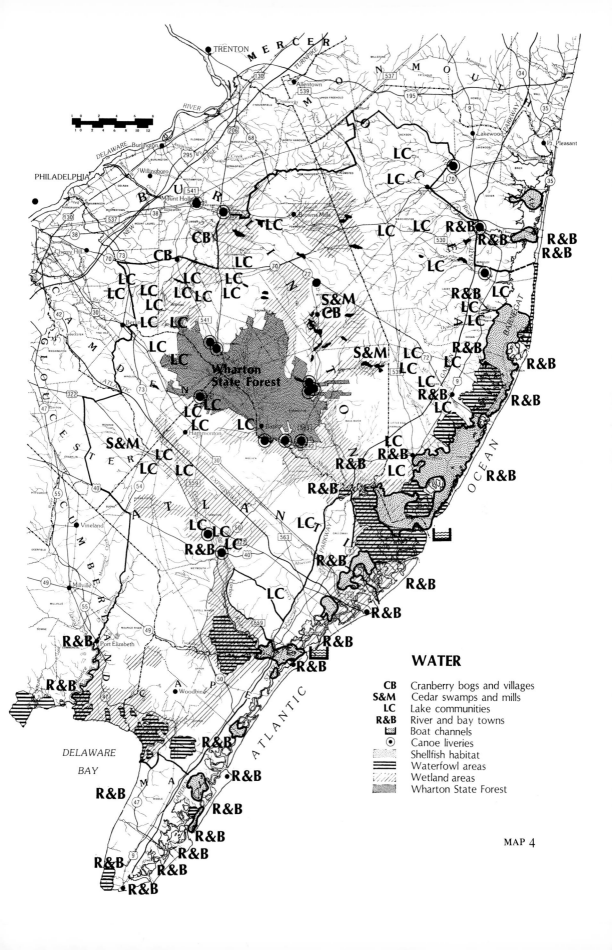

WATER

CB	Cranberry bogs and villages
S&M	Cedar swamps and mills
LC	Lake communities
R&B	River and bay towns
	Boat channels
	Canoe liveries
	Shellfish habitat
	Waterfowl areas
	Wetland areas
	Wharton State Forest

MAP 4

valley to shallow valley. The influential seasonal water-table fluctuations dictate the amount of water in the sandy soil profile, thereby causing differences in soil properties and potential human use. Vertical water movement is, however, overshadowed by horizontal underground flow. In the Barrens, surface water is groundwater because water stored underground accounts for almost 90 percent of stream flow (Robinson 1980). Even in times of extreme drought the flow of water from underground sands and gravels continues to replenish the streams that flow to tidal rivers that empty into estuaries, bays, and the ocean. Groundwater also finds its way from the middle of the uplands to the middle of the bay via pervasive underground flow that discharges out through the salt marshes. The mixing of the saltwater and freshwater in these areas provides numerous habitats for fish, shellfish, birds, turtles, and fur-bearing animals.

The interconnections of the entire Pinelands system cannot be overemphasized. Someone earning a living from the waters of the Pines can rely on a renewable resource, but that resource is vulnerable to disruption. Although the landscape has distinct valleys, watersheds are connected by interbasin transfers of water. Two rivers flowing in different directions but rising in the same region are connected underground. A sandy wedge of sediments provides intake areas for these underground connections near topographic divides where there is at least ten feet to the water table. The natural sumps, or drains, allow deep penetration of water which, when it goes deep, cannot quickly surface in the nearest marsh or stream, but continues underground and reaches the surface some miles downstream (Robinson 1980) (photo 2). People many miles apart and in different watersheds are, thus, irrevocably linked.

PHOTO 2. *A Headwater in the Oswego Watershed*   ©James F. Gandy, Jr.

Water quality in Pinelands streams, lakes, and groundwater is generally excellent. Only along the coast has saltwater intrusion and pollution from septic tanks and sewers forced people to abandon their water. In small areas, these disruptions have been severe, not only because people throughout history have relied upon the profitable and intensive use of the acidic fresh streams and bogs, the fertile brackish estuaries, and salty bays, but because wells throughout are fed by water from the uplands.

The resource may be vulnerable, but at least there is plenty of it. Rain and snowfall generally occur in steady and dependable patterns because of the region's humid coastal climate. Precipitation continually recharges water lost from the cushion through wells and movement into the ocean. The sandy cushion is over two thousand feet thick, descending almost a mile down under Atlantic City. The top of the cushion, exposed throughout most of the Pine Barrens, is the Cohansey sand, which holds over seventeen trillion gallons of water (Rhodehamel 1970). Below the Cohansey lie other layers of sand and gravel, most notably the Kirkwood and Raritan Magothy. Since there is plenty of groundwater, there is plenty of stream flow. Pinelands watersheds are generally small, but their annual discharges are greater than basins of comparable size in similar climates (Robinson 1980).

Large parts of the Pine Barrens are wet. Within these areas are found a remarkable variety of users whose economic and emotional lives are tied to their use and control of water resources. Each of these users has struck a balance between his own needs and the opportunities and constraints of the wetlands and open waters. Understand these balancing acts and one can understand the continuing and evolving historical relationships among people, their cultures, their institutions, and their emotional investment in water resources.

Seasonality of use is a strong pulse in the life of the Barrens. It stirs memories of the past, and year after year brings people in contact with the waters. To some the change in seasons means a time to switch from one resource to another; to others it signals the time to manipulate water levels; and finally to outsiders it determines the only time they come to the Barrens to canoe the rivers.

In the Pines the traditional system of resource use is to deplete one resource until the signal comes to switch to another. Tradition and love of place join with necessity to make people guard the health of the clams, the muskrats, and the snapping turtles, to know when it is time to leave one alone and use another. Those who live along the shore and use the brackish waters for a livelihood know the signals from the environment and the marketplace. Their systems are tied to survival, to enjoyment, and to memory, and it is a rare balance in the modern, overexploitative world. The baymen of the Atlantic coast have one foot in the bay and the other in the woods. Just as water from the uplands ties the forest to the estuary, so does the bayman, from his settlement at head of tide, go down into the marshes and bays and up into the forest to earn a living and balance his use of place.

The bayman harvests from the waters but does not manipulate their flow. The cranberry grower seasonally harvests the domesticated berry and must control water levels throughout the year. Cranberry cultivation needs an intricate system of water works, a skilled labor force, and an aggressive and controlled system of marketing. Even more important, cranberry cultivation demands control of land and water. For every acre of lowland bog, growers feel vulnerable unless they own ten acres of watershed. Such demands for land have created family empires in the Pines, like the

Haines's in Hog Wallow, and a pervasive private property ethic supported by the control of boards that make decisions about municipal land-use and state agricultural policy. Such an ethic may not heed disruption of the delicate water-supply system on the edges of cranberry lands because the major core, controlled by a few, will be protected by private ownership. It takes a special kind of dedication combined with a sense of the past and a vision of the future to be the heart or conscience of the community like Mary-Ann Thompson and Bill Haines, whom we meet later. Every healthy society must have visionaries who will oppose friends and relatives to champion the common interest of the place—to seek a new balance or to redress the old one. This is not just an internal struggle, for outsiders also play a major role.

Use of the waters and, thus, control of the Pines is a tug-of-war between outsiders and insiders. Throughout the history of the region, outsiders have exercised a disproportionate degree of control over the landscape. Water resources served, first, as the base for the pioneer industries; second, as the essential environment for bog iron mining; third, as the potential metropolitan water supply; and fourth, as an attractive resource for suburban settlement and seasonal recreation. Finally, northern New Jersey, in drought years, has looked longingly to Pine Barrens aquifers for relief. Although water cannot be exported under present Pinelands Commission regulations, pressure to export it will remain until alternate water resources are found.

To trace the flow of water is to follow the flow of influence, emotion, immigration, and cultural change and stability in the Pines. Observe the coastal marshes and bays and one will see a centuries-old pattern of balanced resource use. Observe the flow of groundwater to the streams and into the estuaries and one will see the battle to preserve the upland water supply and the constant interplay between outsider and insider. Know these people and their balancing acts and one will know much about the Pine Barrens's major resource—its water.

Brackish and Saltwater:
The Collective Memory and Seasonal Cycle

CLAMMING AND GUNNING

At the age of twenty-two Don Zehner had already spent almost half his life as a bayman on the coast of Atlantic County adjacent to the Pine Barrens (fig. 1). Like all baymen—or "proggers," as they are sometimes called on Delaware Bay and the Eastern Shore—Don has an impressive knowledge of what happens in brackish and saltwaters; he knows how to follow the rhythms of the place. He has the independence of spirit, distrust of authority, and family and community loyalty that mark bay people. To learn about the bays I (J.S.) got into the habit of going out with him, among other baymen, several times a year.

A series of rhythms govern the activities of people like Don Zehner; some rhythms are predictable, some not, some are physical, some biological, and others social. To make decisions baymen have to be aware of these, at least on some level. Of all physical and biological rhythms, the most predictable and shortest are the tides that sweep back and forth across the salt marsh and bays, in and out of the inlets, and up and down the tidal rivers, ranging on an average between four and six feet from low to high tide. Tides tell a clammer where he can or cannot clam, and, since fish and crabs move with them, the tides provide signals to fishermen.

Seasons are predictable, at least to the extent that they change. When they will appear or depart, however, varies by several weeks from year to year. Seasonal changes bring changes in activities. Don clams from spring through fall; he may catch snapping turtles in spring or crab and fish a little during clamming season. Late in fall he hunts ducks and traps muskrats and foxes, and in winter he traps and cuts wood, depending on the weather. The migrations of fish are as predictable as the seasons: cod, mackerel, flounder, weakfish, crabs, and bluefish will all arrive and leave seasonally as water temperatures change. The same is true of waterfowl.

Some physical and biological rhythms are not predictable. As does the arrival of seasons, the climate also fluctuates, so some winters are more bitter than others, some summers hotter or springs drier than others. These differences are important to baymen: While Don may be prevented from trapping in January because of especially heavy ice, he might instead gun for ducks that have been driven into close flocks because so little open water is left. A dry year means less fresh water from rivers entering the estuary, so the normal patterns of salt and brackish water change, and crabs may not penetrate the creeks very far while perch may remain deeper into the estuaries.

Biological cycles are often unpredictable. When I first came to this region, I heard fishermen bemoaning the loss of the great weakfish runs of a decade earlier; they predicted weakies would never return. By the mid-1970s weakfish had returned in enormous numbers, some say even record numbers. No one knows where they had been, nor do baymen know what happened to the striped bass, or rockfish, which disappeared in the sixties and have only sporadically returned. Clam populations are notoriously difficult to predict because some years are good for their breeding or "set," and other years are dreadful. In the early 1970s surf clam populations dropped precipitously, but after the "red tide" summer of 1976, with its subsequent fish kills, surf-clam populations erupted, and clammers are still harvesting the results of that population explosion.

FIGURE I. *Garvey with Clammer and Shinnecock Rake*

The most unpredictable natural cycles are those of the hurricanes and northeast storms, or nor'easters. Anyone who watches the weather knows how notoriously fickle are hurricanes. In the twentieth century, hurricanes have tended to sideswipe the Jersey coast, although some of those swipes have been devastating, such as the hurricanes of '38, '44, '54, and '60. Nor'easters appear from fall through spring on an irregular basis, but it is rare, except in drought years, that they do not arrive sometime during storm season. These storms, which deepen off Cape Hatteras, North Carolina, generally move directly northeast up the coast, bringing with them winds that move in a counterclockwise direction from out of the northeast. The damage nor'easters visit to the coast derives from the power of the wind, its almost unlimited fetch over the ocean (that is, the direction and distance from its origin), and the storm's duration. This last is the crucial factor; a fast-moving storm with high winds might do little damage because the winds do not have time to create powerful waves. But the storm of '62 made up over Cape Hatteras and stalled for three days in March on the Jersey coast. Its average winds were only 36 miles per hour, but for 36 hours it did not move at all. Those winds, given that time and fetch, generated 18-foot waves. In addition, the sun and moon were at their closest point to the earth at that moment, which created perigean tides—tides that went almost 11½ feet above average high tide. The storm destroyed pieces of islands; it picked up parts of houses on the beaches and hurled them back into houses behind them; it rearranged dunes and channels, sweeping sandbars away and depositing the sand elsewhere. Everyone who then lived on the coast remembers that storm. Bay people have to expect storms, in order to know when to head for safe harbor; they have to expect channels and spits to migrate. They must be ready to search for clams in different spots from one year to the next.

As if physical and biological cycles were not enough to keep Don Zehner occupied, he also had to understand economic and political changes, neither of which are as generally predictable as the tides, or even the seasons. Don, like other baymen, is an acute observer of the local and regional markets, otherwise he would not be able to make a reasonable living. He switches from one resource to another as the market dictates. Public policy may also change Don's activities, or at least modify them. For example, the decision of the federal government to clean up America's waterways and fund regional sewers has opened many heretofore condemned shellfish areas on the coast and changed clamming patterns. Of even greater importance was New Jersey's decision to preserve coastal wetlands, thus preserving its shellfisheries.

To understand how a bayman like Don Zehner handles all these variables, we will follow his activities and describe how he sees his place. Don generally spends three-quarters of his working hours clamming, but in 1978 his clamming season started the last week in March, a full month later than usual.

I went out clamming with Don that March morning when the temperature was still 28 degrees, the warmest morning since early December. The killing winter of 1977–78 had iced in the bays, killed the railbirds, the muskrats, the brant, and the black ducks, and driven the snow geese to South Carolina. Winter had frozen in the boats and the docks, had destroyed wharves and the carefully laid plans of managers of the Brigantine Wildlife Refuge who had hoped to overwinter, as usual, the tens of thousands of migrating and wintering waterfowl.

It was 5:30 in the morning when I arrived at Don's house, in which he lived with his parents and sister in a sparsely populated suburban forest setting about ten miles inland from Atlantic City. He usually did not clam in such cold weather, but he needed the money to continue to attend college and to add to his savings for his upcoming marriage. I found him pulling on his long johns, then his flannel shirt and tan work pants, his red suspenders and work socks. We had a quick cup of coffee and then went to the mud room to get his gear—boots, tongs, clam rakes, and baskets that had been waiting from the week before when he planned to go clamming, but thought better of it when a nor'easter threatened to blow up.

We took the gear out, threw it in the back of his pickup, and drove six miles out to Oyster Creek at the end of Leeds Point in northern Atlantic County near the mouth of the Mullica River. The name Oyster Creek identified not just the creek, but the place on it where the inn and boat docks were located; place names like that are common along the Jersey coast—Bass River, Tuckahoe, Toms River, Shrewsbury.

It was still dark when we got to Don's boat slip, which was reached by walking over planking supported by cedar piling driven into the salt marsh. We made two trips to his eighteen-foot garvey carrying first his clam rakes and tongs, then two clam baskets and a fresh tank of gas (fig. 1). He had an old cedar garvey, the classic shellfishing boat of coastal Jersey waters, a broad-beamed, high-gunwale boat that serves as a floating platform for clammers. We headed slowly out of the creek with the falling tide. The air was still; dawn had barely cracked the horizon, and snow geese began their gabbling as they lifted off their night resting places to head for the few cordgrass feeding spots left open by the hard winter. Two flocks had just returned from the Carolinas. At least the harsh winter had given the cordgrass marshes a rest from the depredations of the geese. On the decline for fifty years, the snow-goose population had rebounded with a vengeance during the 1970s when hunting was prohibited and DDT banned. Snow-goose hunting was renewed in 1975, but goose populations continued to increase above 50,000. One of their favorite wintering grounds was in and around Brigantine Wildlife Refuge, and they were eating themselves out of cordgrass, which they pulled up by the roots, leaving mud flats that tides and storms eroded. The snow geese were slowly destroying their own habitat. Don Zehner noticed these fluctuations because he himself used every resource of that bay and river.

Waterfowl gunning is as much a part of the coastal waters as is clamming. There are stories from the turn of the century of gunners coming home with 200 birds apiece. Market gunners could clam and fish in the summer, and earn their cash in fall and winter selling duck and goose meat to the New York and Philadelphia markets. In those days duck was cheaper than beef. Gunning was partially responsible for the drastic decline in waterfowl populations because hunters developed enormous shotguns, like howitzers, which could wipe out a score or more of ducks sitting on the water. Equally implicated was the destruction of nesting habitats in northern marshes from New York State through the Canadian Arctic.

Thousands of waterfowl hunters now gun the Atlantic coastal marshes annually. In fact these hunters account for a surprisingly small percentage of waterfowl mortality. On the Brigantine Wildlife Refuge, for example, more than 100,000 waterfowl visited the area, most of them snow geese and black ducks. Hunters killed only 3 percent of the snow-goose and black-duck populations in 1980 according to New Jersey Department of Environmental Protection reports. Waterfowl hunters have nothing in common save their sometimes fanatic devotion to duck hunting. People who hunt waterfowl tend to do so to the exclusion of almost all other hunting. They have their own national organization—Ducks Unlimited—and spend a great deal of time and effort on their equipment, which must include a good duck boat and motor, a series of blinds from which to hunt, two or more shotguns plus steel shot (lead leaves poisonous residues and is banned on coastal marshes), and two dozen or more decoys (also called dekes or stools) which are carried on board and carefully set according to prearranged patterns near the blind. Waterfowlers come from all backgrounds, poor and rich, from baymen to machinists, doctors, lawyers, and politicians. Waterfowl hunting is as important a recreational activity in the coastal section as is any other, and duck hunters form a powerful conservation lobby.

The Jersey coastal wintering grounds for waterfowl are as productive, if not as extensive, as any along the East Coast because of the richness of the wetlands—the cordgrass or salt hay marshes and eel grass and sea lettuce that are fed by the tides, the rivers, and the decaying remnants from each annual growth. The detritus feeds the algae and plankton, which feed the shrimp, clams, crabs, and fish fry, which feed the fin fish. Different ducks select different delicacies: to the teal and brant go the sedges, reeds, and grasses; to the merganser and scoter go the fish and crustaceans; and to the black duck goes almost anything available, although blacks prefer vegetation. A good gunner knows the condition of the winter habitat, knows what food is available, and so knows whether it is worth shooting black ducks that day or not. Most baymen prefer to eat fish rather than a fishy duck.

November is the busy month for waterfowl on the Jersey coast, for that is when populations reach their peak. Besides the more than 50,000 snow geese, there are 10,000 to 20,000 brant and an uncounted number of Canada geese, mallards, black ducks, teal, shovelers, scaup, buffleheads, scoters, mergansers, and coots with a smattering of whistling and mute swans, gadwalls, baldpates, pintails, wood ducks, redheads, canvasbacks, goldeneyes, and ruddy ducks. The richness of the marshes can be seen in the wealth of ducks alone during November. After November some ducks stay, others continue south. The spring migration is slower; mortality has been high from winter, disease, and predation, and the waterfowl are often in such a rush to reach their nesting areas that many pass by their winter resting place.

I had accompanied Don Zehner on some of his best gunning days, the bitter cold mornings of December when ice began to form on the bays, driving ducks into tighter flocks as the amount of open water diminished. A perfect day was a bit windy, cold, and overcast with a low cloud cover that forced the ducks and geese to fly low, giving them less time to inspect a hunter's blind and yet not preventing them from flying throughout the day, as would a high wind. On such days Zehner left behind the bulky garvey, put a little four-horse engine on his sneakbox, and set off down Oyster Creek. The sneakbox (fig. 2), developed during the nineteenth century, is a marvel of adaptation specifically designed for gunning on Jersey salt marshes. Generally twelve to fourteen feet, it has a very shallow draft, with room for strings of decoys, and it can be camouflaged for use as a mobile blind as well as for sneaking through marshgrass at high tide.

Instead of going to the clamming areas of Great Bay, however, Zehner would stay within the maze of ditches and riverlets which cut through the marshes, most of which lie within the federally owned Brigantine Wildlife Refuge. Sometimes Zehner crossed the bay to his snapper-turtle territories, on which he has permission to hunt either by custom or lease. A good morning for Zehner would be a limit of ten ducks, and because he did not need the meat, he rarely killed more than his limit. Half the joy in any case, he said, was to be on the water at sunrise listening to the sounds and watching the flights. These were mornings of quiet enjoyment. Gone were the opening days of October when the ''bozos'' with bright clothes and raucus voices squealed on their Orvis duck calls and blasted aimlessly at huge white whistling swans far out of range on the open bays.

Don had not shot many geese or ducks that winter because of the cold weather, but he recalled the times and places he had had good shooting days as the garvey made its way downriver that March morning. At age twenty-two, he had already spent part of the last eleven years on the salt marshes and bays, and had memorized the channels and bars along a twenty-five mile stretch of coast. He did not need much light to get out to his clamming area. He was following the memory of his childhood, of his father and the baymen who settled the Jersey coast in the late seventeenth century.

The collective memory of generations of bay and river people contains powerful information. The information comes partly from the symbols of shipwrecks, whaling, Eric Mullica and other first settlers, and from a love of the seasons of marshes and rivers that ties residents to their landscapes. But more than symbolic power exists in the memory, because the information it contains is part of Zehner's and all bay-people's livelihood—the knowledge, for example, of tides, winds, and storms that open and close inlets, shift sandbars, create and destroy clam beds, and wreck boats (photo 3).

Zehner knew where the deeps and shallows were, the best clamming and crabbing grounds, and when to stay out on the bay and when to head home in the face of a shifting wind that presaged a storm. His knowledge came from understanding elemental processes; he knew which way and how far sandbars were moving and, depending on how severe the storms were during the winter, whether channels that hold flounder would still be where they were in the summer.

FIGURE 2. *Sneakbox with Hunter and Decoys*

PHOTO 3. *Aerial View of Great Bay and Little Egg Inlet*

Zehner, like all baymen, saw flux where others saw stasis. The proponents of stasis—homeowners, real-estate developers, bankers, and federal flood-insurance supporters—consider the beaches and barrier islands their property when, in fact, the sea owns this area. Periodic nor'easters and hurricanes will continue to claim the beaches and anything on them. Older generations, knowing this, built their communities on the mainland and never erected permanent structures on islands and marshlands.

Zehner was accustomed to the flux, and headed out into Great Bay to search for a clam bed he suspected had been in a sufficiently protected area to have escaped winter destruction. The clam bed was a sandbar in the eastern end of Great Bay near the old fish factory, abandoned in the fifties; the factory, one of a string of similar factories from Long Island to Florida, had reduced menhaden, an anchovylike fish, to poultry feed and fertilizer, and in its time had reduced bay people to tears when the fierce smell of processed menhaden was driven over fishing boats and to the mainland by prevailing southeasterlies in summer. Local people considered it a mixed blessing when the fish factory closed.

Zehner headed the garvey's bow straight toward the clam bed. Tide was not yet dead low and, unloaded, the garvey drew only eleven inches of water, so he did not worry about running aground on the sandbar. He knew where most of the shoals lay anyway, partly through experience and partly because he read the water well, the telltale sign of fast-running water where it moved through a trench alongside a shoal, and the water's color, which is lighter in more shallow areas. Zehner's route took him through what appeared to be a forest of dead cedar trees growing in the northern section of Great Bay. Had the cedars not been planted so neatly in rows, the forest would have looked just like the remains of cedars drowned by a beaver dam, their trunks stuck three or four feet into the muck and their leafless branches flat against the dawn sky.

The trees of the forest belonged to clammers who leased clam "lots" or parcels of the bay from the state of New Jersey, which owned the territory. State biologists for the Division of Fish, Game, and Shellfisheries selected areas of moderate-to-low shellfish production for transplant or relay leasing. Highly productive sites were left open to public shellfishing. The relay program allowed commercial clammers to take clams from polluted waters and place them on the lease lots for thirty days to cleanse themselves, after which they could be certified as clean and sold like any other clams.

Zehner was one of more than 30,000 licensed clammers along the Jersey coast, most of whom were senior citizens or recreational clammers who were allowed no more than 150 clams a day. Most clammers and clams were found just north of the area Zehner worked, in Barnegat Bay. Zehner worked his relay lot heavily in the summer, but like most commercial clammers, he generally clammed in public waters. Although commercial clammers accounted for only about 8 percent of the total number of licenses, they took 60 percent of the clams. Still, for Zehner, the surest way he could make money that year was to start with his relay lot in Great Bay.

Zehner had one lot in Great Bay on which he renewed his lease annually. The lots cost $50 an acre and were granted by the state's Shellfish Council, a group of the governor's appointees. Once a lease was granted, it generally stayed in the family or among friends, and the Shellfish Council could not take back a lease except by complex legal maneuvers. The council could, however, grant new leases to whomever it chose, although new leases were rarely granted. In general, the council acted to grant leases, extend or diminish seasons, open or close waters to fishing, and limit catches on the advice of fisheries biologists.

That March morning Zehner skipped the garvey through the cedar stakes, past his lease lot out toward his clam bed. He slowed the boat as he approached the shoal, turned off the motor, threw out a small anchor in about three feet of water, and picked up his tongs, a pair of five-foot crossed poles with rakes on either end. When he closed the top of the poles together, the bottom ends with the rakes also closed, trapping large pieces of material and allowing sand, gravel, and small clams through. Tongs were most often used in gravels and hard bottoms in which leverage was required to scrape the clams up, and Zehner remembered there was some hard bottom at the north end of the shoal where he had started clamming. He worked at an even pace, letting the tongs do as much of the work as possible. The wide, flat bottom of the garvey served as a stable platform from which to work. He picked up more than fifty clams in the first half hour, not as many as he had hoped for. With the price of clams at $100 per thousand, it would take him five hours to make his goal of $50 and the tide would be rising hard by the time he finished. Zehner put the tongs away and moved the boat a little south toward the sandy areas where he could use his Shinnecock rake, a long pole with long tines and a basket on the end of it. He was now in only a foot and a half of water, and tide was almost dead low. He began raking in half again as many clams as he had before.

Even if the water had been warmer than 45 degrees, Zehner would not likely have gone overboard to rake clams. He simply did not like wading. "I ain't goin' in that water," he said. "There's monsters in there." Many clammers, on the other hand, preferred to wade at low tide because it made the work easier. One could use the density of the water to help hold up the clams instead of muscling them with less leverage straight into the boat. Some of the old clammers, when the water warmed up in June, would almost exclusively "leg" clams by searching them out with their booted feet, kicking the clams up into the water, and catching them with their hands. A good legger could work as quickly as a wader in a productive area, but the art of legging was beginning to disappear. All his life Don Zehner would remain a boater and sometime wader.

At the end of four hours Zehner had his five hundred clams. As the tide began to run in, he started his motor and went with the tide back to Oyster Creek. He wanted to drop off the clams at the wholesaler in Pleasantville, pick up his money, get home to shower and change, and be at school for a 12:30 class.

Heading home, Zehner passed a party boat going out toward the ocean, winding first south, then north to avoid the sandbars, and through Shooting Thorofare, which is the last section of Little Egg Inlet before one reaches the ocean. It was Jimmy Kelson from Port Republic who also kept his boat at Oyster Creek; he was taking a party of six to the cod grounds. Because winter had been so long, water temperatures stayed cold, and the cod stayed in larger numbers than would have normally been the case. Kelson had once been a commercial fisherman, but his age and especially the poor economics of commercial fishing on the Jersey coast had led him to a head boat operation; that is, he took people out fishing at so much per head, usually about $40 a person for a day's fishing. At that price he knew people expected to catch fish, and he made sure they did. Kelson catered more to local people or outsiders who had been visiting the area for some years than did large boats that worked out of Cape May, Atlantic City, Mana-hawkin, or Toms River. The large boats could take up to seventy-five people and often did in the summer when out-of-towners arrived in full force, ready for a boat party when they stood on the deck of a large party boat, rolling with the swells, a fishing pole in one hand, a beer in the other, alternately guzzling beer and vomiting it back overboard, twisting each other's lines up in a tangle and yelling, "I got one, I got one!" Kelson did not care for such scenes, lucrative as that business could be. In fact, there was an undercurrent of resentment on the part of clammers and small party-boat owners toward the big operators. Zehner and Kelson preferred to keep the number of tourists to a minimum, while the folks at Cap'n Starns in Atlantic City could not get enough of them. At bottom it was a conflict of life-styles that would be played out innumerable times with different backdrops. Many locals preferred the slow way of life, while others welcomed a faster-paced economy with its attendant suburban and casino development.

Without knowing it, Don Zehner was trying to bridge a series of gaps between seasonal life on coastal waters and encroaching development. Not that old-timers were immune to outside pressure. Their cash economy was tied to regional markets, they had worked from time to time for corporations or government agencies like the county highway department, and they hired out to tourists who needed guides for their local products. Old-timers had to avoid clamming in polluted waters throughout the twentieth century, and they also had to protect their crab traps from out-of-town poachers. The problem was less a matter of qualitative than quantitative change, but after 1960 the intensity of pressure to change from rural seasonal to suburban service economies increased enormously. Coastal landscapes in the north transformed themselves as Ocean County became the fastest-growing county in New Jersey from 1960 to 1980. Bay people's communities, infiltrated by metropolites, changed character. Old residents knew that they were losing their grip; the centers of their lives no longer held together. Their towns were no longer their own. Gladys Eayre, a sixty-year-old resident of Waretown, whose family came to the coast in the eighteenth century, could no longer walk unhindered along the marshes at dawn. Not long ago a policeman stopped her for loitering at 6:00 A.M. by the marsh and shadowed her all the way to her house in his squad car. Merce Ridgway cannot keep roosters and hens in his backyard, as roosters are banned in Waretown. Ridgway has threatened to get a bunch of guinea fowl which scream instead of crow at dawn. Thousands of similar stories exist.

What is so valuable about the old ways, the seasonal activities, and family and community ties? Surely, intrinsic interest is part of the value—the old way of doing things, crafts, folktales, and work. More important is that seasonal life-styles reflect strong and mutually advantageous balances among people and their resources. These balances evolve over time and mediate changes in the supply of food, water, shelter, and living space as human demand rises or falls. They have to do with subsistence, with survival without dependence on highly technological, capital-intensive processes. But the survival must be seen in human terms. People along the coast, indeed throughout the Pine Barrens, adapted themselves to live not only with their resources, but interdependently. Suburbanization and industrialization tend to break apart carefully constructed ways of doing things, of getting along. Once broken, the old ways are hard to piece together, and a whole series of adaptations fall apart.

For every resource available on the coast to newly arrived Europeans or New Englanders there was, or would soon be, a method to use the resource—from salt hay to herons. Don Zehner and others like him were heirs to all the skill that had grown over thousands of years, and the richness of the coastal resources reflected the richness of strategies to extract them.

"We may be poor, but we never lack for friends or food," is a Pine Barrens aphorism. To every season, its resources: summer and spring had snappering, fishing, clamming, crabbing, oystering, and salt haying; fall had all these plus waterfowl hunting; winter had some fishing, but mostly trapping, upland hunting, and timber cutting. There were always choices; if one resource were depleted or inaccessible, or unmarketable, there was another to exploit.

CRABBING

Most clammers and watermen crab from time to time depending on the supply of crabs, the market, and their own schedules. Unlike crabbing on the Chesapeake, where it is a large-scale, full-time operation, the crab industry on the Jersey coast and Delaware Bay is limited to part-time operators who are reluctant to devote all their energies to less extensive, less promising waters. In fact, most of the crabs one buys at fish stores along the Jersey coast are shipped from the Chesapeake. (The best study on crabs and crabbers is William Warner's book on the Chesapeake, *Beautiful Swimmers*.) There are a few crabbers and clammers who provide for the retail stores along the coast, but the market is generally so strong from late May through September that they can sell their crabs at retail if they have small keeping pens.

No one knows how many people crab on the coast, and no one has any idea how many crabs people harvest during the season. The state of New Jersey, until 1982, never required any kind of license, so no records exist. Beginning in 1982, the state required all commercial crabbers, or people who sold crabs they caught, to pay a $100 fee for a license, and subsequent surveys will determine how many crabs can be harvested. At one point the state wanted to license any person with more than two commercial pots, but recreational crabbers from the whole mid-Atlantic area raised such a cry that the idea was dropped. Crabbing is one of the region's most important summer recreational activities, but it will probably be five years before anyone knows how important.

Ted Von Bosse, who lives in Port Republic, Atlantic County, is such a crabber. He is, in his unusual ways, representative of part-time crabbers from Toms River to Maurice River. Ted, when asked his occupation, replies that he is a college teacher and that activity engages him for much of ten months a year. He also runs a small boatyard, carves decoys, sometimes clams, fishes, and hunts waterfowl, and, as a certified public accountant, handles a small number of accounts and occasionally testifies before congressional subcommittees on economic issues. One summer he and his wife, who is a carver, built the log house in which they now live.

In summer, regardless of weather, Ted goes crabbing on the morning or afternoon dead tide, low if possible, but high tide if necessary. He puts on his bootless yellow suspender oilskins and, with his heavy orange Day-Glo rubber gloves in hand, heads toward his sixteen-foot garvey. He bought the old garvey for $35 from one of his ex-students, and the fiberglass cedar boat has been seaworthy, or at least bayworthy, for the past ten years. He bought the eighteen-horse Johnson motor from a fellow town-council member for $75 (a brand new one costs about $1,000), and the motor has never given him trouble. The only thing he has added to the boat is a small bait-cutting table two-thirds of the way back from the bow on which rest the crab pots when he pulls them from the water. There is still room for his ten-year-old Chesapeake bitch, Buffy, who helps with the harvest.

With him Ted takes three bushel baskets (a dozen for $15 bought in Hammonton) and two five-gallon containers of fish skeletons and heads for bait. He collects most of his bait from incoming fishermen who use his docks. The bait needs to be fresh so it can freshly decay in the water; Von Bosse's big old freezer at the yard is always crammed full of fish remains, which he generally has to pry out of the freezer with a crowbar. Oily fish like mackerel and bluefish make the best bait, but anything will do—flounder, weakfish, croaker, perch, sea bass, rockfish, whatever.

Ted generally has from twenty to thirty-five traps out, most of them in Nacote Creek (photo 4). He has fooled with eight or ten different kinds of pots, as crab traps are called, and continues to experiment a bit, but has basically settled on two types—the $8 Eastern Shore (Chincoteague) type made from thin galvanized-steel mesh that lasts two years before rusting out, and the $17 Freeport, Maine, heavy-duty, thin-wire model

PHOTO 4. *Crab Pots and Eel Pots*

that can last eight or more years. Von Bosse says that trap losses due to thievery and boating mishaps are high enough to cut the expected life of an expensive trap in half; this being the case, most of his traps are of the $8 variety. He is trying out an $11, long-lasting, vinyl-covered wire model made in the Chesapeake that a friend uses extensively, but Ted does not like it. He says the mesh is too thick and scares crabs away from the pot, so he claims he has had less luck with it. Ted could make his own traps, but it is not worth his while. He would rather spend the time carving decoys.

To warn errant recreational boaters of the location of crab pots, the state of New Jersey until 1981 required crabbers to put a cedar stake, like clam or oyster lease stakes, where traps exist. Ted's stakes are tipped in orange, the color of his gloves, and are pushed by hand two to three feet into the mud. On each stake is attached an iron ring, and attached to the ring is a line on the end of which is the pot. At high tide the pots lie from three to eight feet down, waiting for the crabs that wander ceaselessly up and down the river in search of food.

Vacationers and weekenders often know what the stakes mean, and crab pot thievery is a common problem. There are stories of outsiders trailing crabbers or clammers to find their most productive sites or pots and then returning to steal the crabs or clams. If this happens more than once or twice in a month, Ted, as will other crabbers, takes time to track down the culprits. He or his son, or a friend in a boat other than the garvey, will patrol the water until the thief is caught. Ted gives the thief a choice—$50 plus the retail price of the crabs or a summons by the marine police. The money and, especially, the fright in being caught, are enough to end the matter. Once in a great while a local boy will be caught stealing crabs or clams. The punishment is, in these cases, swift and harsh, and the local will soon find his boat stove in.

Ted stops at his first stake, turns off the motor, grabs the stake to hold the boat steady, and takes a loop of rope from the boat. He doubles the loop around the pole, through itself, and loops the remaining end around a cleat on the gunwhale to tie the garvey to the stake. He then takes his home-made boat hook and begins searching the mud bottom for the pot line. Up comes the pot, which he grabs with his bare right hand while his gloved left hand pushes two crabs off the side toward the bottom of the pot. A crab pot is a most cunning device. At first glance it looks like a simple square wire cage, but on inspection inside it has funnels and compartments. The two to four funnels in one section lead to another cylinder in the center of that section which holds the bait. The crabs enter the funnels and, generally unable to find the narrow end of the entrance, stay in the trap. Ted first opens the bait cylinder, throws out the decayed bait, replaces it with fresh, closes the bait door, and shakes the crabs still in that section down to another section. He then flips the pot over and shakes the crabs to the last section, flips the pot a last time, and unhinges that section into which the crabs have by now come to a dead end. Working quickly, Ted's gloved hand picks up the crabs and throws them overboard if they are small or into the bushel basket if they are edible.

Occasionally, a crab manages to pinch the orange Day-Glo hand, and Ted tells the crab to cut it out. He often talks to the crabs as he puts them in the basket. "This trip," he says to one crab, "is the most important one in your life. You've been waiting for this one, haven't you?" To another, "Well, you're a country crab. Want to go to Ted's crab city, see all your crab cousins? You'll like it there." Some pots yield ten or twelve good crabs, some three or four, and one has none in it. Ted cannot figure it out; he suspects the vinyl-covered wire.

Meanwhile, Buffy, the bitch, stares into each pot as Ted pulls it out of the water. She whines and waits. Finally, one pot yields a small male diamondback terrapin among the crabs. The terrapin is still alive because Ted generally works the area twice a day. The turtle would have suffocated if left in the trap one more tide change. Buffy is clearly excited and, when the gloved hand presents her with the terrapin, she snatches it, jumps overboard, and swims with the turtle in her mouth to the near bank where she gently deposits it, and then swims back to the boat. By then Ted has emptied the pot and hauls the dog back into the boat where she shakes herself and continues her whining wait for the next terrapin.

There are no more terrapins that day. As summer lengthens and prolonged rains lessen, the river becomes more salty and the terrapins move upstream into smaller creeks. The saltier water also means more female crabs will begin appearing in the traps, although 80 percent of the crabs are still males that tend to wander farther astream, especially in less salty waters.

It takes Ted two hours to finish his round of twenty-five traps, by the end of which he has two baskets full, not bad for a half-day, especially since most of the crabs are large. Back at the dock, Ted unloads and carries his crabs to his crab pen—a large, shallow wooden pen with a partition in the middle. He pumps brackish water in a continuous stream up from the river through a long vinyl pipe to keep the crabs alive.

Ted can sell all the live crabs he gets during the summer for $6 a dozen or $35-$40 a bushel. He likes his crabbing. It could never make him rich, but it gives him the chance to go out on the water and talk to his dog and his crabs, and to get paid for it.

TURTLING

Snapping turtles. Snapping turtles are a good deal larger and even less pleasant to deal with than crabs (fig. 3). Snappers are born angry; they appear from their shells as snapping three-inch babies and grow into biting four-foot, fifty-pound, nasty, ridge-backed adults. Snappers do not look for trouble and, even if stepped on in the water, will generally draw in their hook-beaked heads—little comfort to occasional waders who have lost toes while wading.

The spring of 1980 found Don Zehner up the tidal creeks setting out traps for snapping turtles rather than clamming. Clams were plentiful enough that May, but the market in the Delaware Valley region had been flooded by cheap clams from North Carolina, and it was not worth Zehner's time to clam, so he took six snapping-turtle pots out every morning for three weeks on his garvey instead of his rakes and tongs. Snappers were selling for 40 cents a pound to Philadelphia markets, most of the meat going to restaurants for snapper soup. The price of snapper meat had remained quite steady for some years. No one, however, knows how many people worked snapping turtles because no reliable figures are kept.

Snapping turtles spend from April through October moving slowly or lying in wait in mucky bottoms of fresh and brackish streams and ponds looking for detritus or live food. One finds the largest and most abundant snapper populations at "any crick which goes into the woods from the saltmarsh" (photo 5). From October to April snappers hibernate and only travel across land in spring to mate or to find a new habitat.

People catch snappers by trapping them with smelly dead bait in large wire traps called fykes (fig. 4). Snapper season has no official beginning or end, but because snappers will not "pot" or enter a trap after July, the season is effectively May to July.

A year after the cold winter of 1979, the weather, as is its wont on the coast, turned sharply warmer, and Zehner decided late in April to take me (J.S.) in his shallow-draft Boston Whaler up one of the creeks that lead to the woods to set fykes. He set his half-dozen fykes at different spots on two creeks in the afternoon, got up early the next morning, and made his way toward the fykes, which he had carefully placed to allow air

FIGURE 3. *Snapping Turtle*

PHOTO 5. *Wigwam Creek near Great Bay*   ©*James F. Gandy, Jr.*

space at the top of the trap during high tide so the snappers would not drown. The fykes were sufficiently heavy that they would not drift with the tide and current, but just in case, Zehner had provided the traps with innocuous cork bobbers.

In the first fyke Zehner found only the fish heads he had used for bait. He hauled in the trap and moved upstream to the second, which held a small snapper between five and seven pounds, hardly enough salable meat on it, and probably unacceptable to the wholesaler anyway. He opened one end of the trap and let the small turtle swim out. The third yielded nothing, and Zehner was puzzled because the creek in the past had yielded some monsters; but he had not worked it in three years. Something, he said, must have changed in the creek; perhaps the shallow bottom had not sufficiently protected the hibernating snappers the past two record-breaking cold winters, and new populations of turtles had not yet reached size. Maybe the beaver dam several yards up into the woods had caused the stream to become intermittent in the summer at low tide and had driven out what remained of the big snappers. Maybe the water in the creek was still too cold, colder than in nearby creeks. Maybe this, maybe that. "Sometimes you get the bear, sometimes the bear gets you," he always said on such occasions and tucked away the information.

Zehner went back out the creek to another larger one a mile up the Mullica River. The first fyke was a bonanza, a rare double, a thirty-five pound female and a twenty-pound male, the bodies of which barely fit in one end of the trap. As Zehner anchored the boat, the snappers stared at him and snapped ferociously. He sat down for a moment to ease his excitement and calm himself into caution. Snappers generally

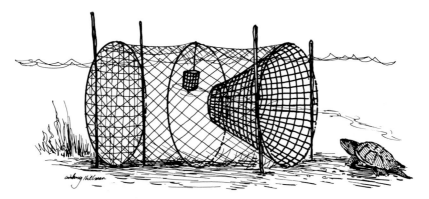

FIGURE 4. *Turtle Fyke*

move slowly, but, like snakes, they can strike out their heads and snap off bones in a fraction of a second or put claw marks deep into one's flesh with their powerful legs. He moved slowly. Fortunately, the big female was wedged into the fyke, almost immobile, with her tail to the door. Zehner gingerly opened the trapdoor with his left hand while he kept the turtles occupied with a stick in his right. He quickly tipped the trap onto the doorless end, flung open the door, grabbed the big snapper's tail, and started twisting the turtle round and round in mid-air to disorient it. Within ten seconds the dizzy snapper was docile enough to allow Zehner to grab a double canvas bag. He twisted the turtle a few more times, stuffed it head first into the bag, and tied the end with strong rope. In the meantime he had kicked the fyke door closed with his foot to prevent the other turtle from escaping. After the big female was secure, he went through the same process with the male, and then he sat down for fifteen minutes to rest.

Zehner finished his morning with one more twenty-pounder, loaded his three sacks into his pickup, and went down to the wholesaler at Pleasantville. He had his snappers weighed, paid the $6 shipping fee to Philadelphia, collected his money, and went home for a rest. He had often thought that if only the market were more steady, he would even spend some of the winter catching turtles—not snappers, but terrapins.

Terrapinning.* There are perhaps no more than a half-dozen baymen along the Jersey coast who still catch diamondback terrapins, the four- to seven-inch, marsh-dwelling turtles which, at the turn of the century, were a standard part of any fancy East Coast wedding feast (fig. 5). Terrapins were so large a part of so many feasts that baymen almost exterminated them, and they have been an endangered species during most of the twentieth century, although they have made a startling return since 1960. Many coastal crossings from mainland to barrier beach now have turtle-crossing signs that warn motorists to watch for females on egg-laying expeditions during springtime.

Although terrapin are now abundant, and New Jersey allows a four-month winter terrapin season, something odd happened on the diamondback's return to its old

*Thanks to Dr. Roger Wood, Professor of Biology, Stockton State College, for information on terrapins and on the late Earl F. Yearicks, with whom Dr. Wood had gone terrapinning several times between 1975 and 1979.

habitats. The taste for its flesh had disappeared from the American palate, and the market for its meat had all but disappeared.

The highwater point for terrapinning in New Jersey was 1897. The recorded catch was down from 13,500 pounds in 1897 to 8,000 in 1901, and only 200 pounds in 1921. Although a sizable catch of 3,000 pounds was recorded in 1929, comparable to the average in the early 1890s, the years from the Great Depression until the present never exceeded 800 pounds. Despite human depredation, however, the diamondback hung on and, because it was no longer eagerly sought after 1930, finally returned in abundance (U.S. National Marine Fisheries Service 1977).

Earl F. Yearicks (pronounced Yerks), originally an oysterman and clammer, became chiefly a ''tarpin'' man at the end of his life. He died in 1979 at eighty-five and terrapinned through the last winter of his life. Earl Yearicks was born and lived his life in and around Cape May Courthouse near the southern tip of New Jersey. He quit school at thirteen, and for the next thirty years worked oyster and clam beds in Delaware Bay and along the Cape May coast. In the late 1940s, just as the terrapins were recovering from near extinction, Yearicks began to gather them as a sideline and ship them to the major East Coast market in Brooklyn, New York.

The terrapin market in New York was steady, but not profitable, and Yearicks merely broke even in 1978 when the price he received for turtles paid for little more than transportation costs, fuel, and wear on his boat and equipment. Once every week or two from mid-November to March a wholesaler stopped by Yearicks's house to pick up four or five baskets filled with twelve to twenty turtles. Yearicks sorted his terrapins into three groups according to size: large six-inch females (measured along the length of the plastron), smaller five- to six-inch females plus four- to five-inch males, and unsalable ones under four inches. In the late 1970s large females fetched $1.50 to $2.00 each and smaller turtles from 50 to 75 cents.

Earl Yearicks was an excellent self-taught field naturalist, as are all baymen, and knew unfailingly where to find terrapins in the twenty-five-square-mile area he worked for thirty years. Diamondbacks are found scattered all over the bays and marshes from

FIGURE 5. *Diamondback Terrapin*

April through October, feeding on a wide variety of invertebrates, vegetation, and detritus. By the middle of November they are in hibernation in the mud of the tidal creeks and marshes until March, a type of hibernation that still puzzles scientists, who cannot explain how terrapins physiologically adapt to the anaerobic environment underneath the mud.

Yearicks discovered that when terrapins hibernate, they choose one of three different habitats. Some simply drop down to the bottom of a tidal creek and rest there all winter; some choose the intertidal zone adjacent to and along creeks and dig in; and others form small groups from six to a dozen individuals and huddle together in the undercuts of tidal creeks, although Yearicks on rare occasions found up to a hundred in large potholes under creek banks.

For each habitat Yearicks had a different technique. He devised a special tool for getting turtles scattered along creek bottoms but not dug in. This was a twelve-foot pole, welded together from two six-footers, with a bicycle basket attached to the end of it and a clam-rake head on the end of the open bicycle basket. He fixed copper tubing onto every other tine of the rake to prevent the sharp tines from puncturing the turtles' shells. He would then drop anchor in a creek and sweep the bottom of likely spots with the motion of a clam raker.

To capture terrapins dug into the mud of the intertidal zone, he would first probe the area with a short rake handle until he heard the right kind of thunk and would then use a three-pronged, hard-tined rake to dig out the turtle. To find groups of them, Yearicks waded the marsh at low tide and simply poked at holes and undercuts until he found a group, which he could then collect by hand.

At the end of a day Yearicks would drive home with his baskets of terrapins, sort them into three groups, and, after covering them with wet burlap, leave them in his garage where they would continue to hibernate. He would keep undersized turtles all winter if necessary, and release them back to the marshes only when they came out of hibernation with warmer March weather.

Over a thirty-year period, Earl Yearicks took more than thirty thousand diamondbacks from the same twenty-five square-mile area while the turtle population continued to increase. Surely, if terrapin ever returns to the American diet, terrapinning will again become a viable option for baymen.

Oddly, Yearicks rarely did any trapping, which comes during the same season as terrapinning. Furthermore, muskrats are often found in the same areas as diamondbacks. Maybe he did not like the idea of killing fur-bearing animals. In any case, while Yearicks was probing for turtles in the Cape May mud, Don Zehner was on the Mullica marshes shooting waterfowl and checking rat traps.

TRAPPING

Don Zehner's trapping season blended into gunning season in December. The same cold that sent waterfowl down from the north also made the undercoats of small mammals grow thick. Whenever Zehner went gunning after the opening of fur season on December 1, he might also check his trapping area after the morning flights. As a teenager he used to trap almost every day all season to March 15 regardless of market conditions. He started muskrat trapping at age fourteen when rats brought $2.50 for brown pelts and $3.00 for blacks ones (fig. 6). In 1979 the furs sold for $7.00 to $8.00 a piece, but Zehner during that year only set his traps in December and February because

FIGURE 6. *Muskrat*

he found woodcutting at $75.00 a cord and fox trapping at $70.00 a skin more profitable.

Teenagers, in any case, trap many more days than people over twenty-five—perhaps because older men get permanent jobs or perhaps running traplines every January morning in zero or fifteen-degree weather can only be done by adolescents testing themselves against the elements. Zehner was typical of the 1,200 trappers in the Pine Barrens in terms of his age, the number of days he spent trapping, the types of traps he used, and the money he made. And, like most Pine Barrens trappers, he spent most of his time on muskrats; in New Jersey muskrats outnumber all other furs combined by more than five and a half to one (N.J. Dept. of Environmental Protection 1978).

In 1979 Zehner leased from a family friend three hundred acres of marshland in a brackish area of the Mullica River. Because the price of skins is so high, competition for good lease areas is heavy, and it is almost impossible for a newcomer to find good trapping territory. Zehner's lease cost $200, a small price compared to the $100 an acre that the most productive Delaware Bay area commands. Zehner briefly considered bidding on land in the Brigantine Wildlife Refuge where he regularly leased a gunning blind, but did not consider it worthwhile to trap there because so much of the marsh is too salty to support large rat populations. Muskrats favor brackish and freshwater zones. Folk wisdom states that where one finds three-corner grass, there will be a lot of rats. Around the Delaware Bay marshes, snow geese compete with muskrats for food, and trappers in Cumberland and Salem counties complain bitterly that, as snow-goose populations have risen and the geese have begun to destroy marshland, rat populations have declined. In the Mullica area, on the other hand, geese tend to stay nearer the coast and the safety of the wildlife refuge.

In 1979 Zehner and a friend set almost a hundred traps during the first week of December. Like almost all trappers, he used conibears, which are simple, devastating spring traps that break muskrat necks. A conibear consists of a heavy spring, attached to which are two five-inch squares of steel (fig. 7). The trapper sets the conibear by pressing on the spring, reversing the two steel squares, and setting a catch to which is attached a two-pronged light trigger. The goal is to set the trap right in a muskrat run and, when the rat travels down the run, it trips the trigger and is killed immediately. A few trappers still use the old leg-hold or jaw trap—the one pictured in cartoons with jagged jaws to catch bear—only these are small and have rounded jaws. While cheap and easy to set, such traps are less efficient for muskrat because a rat can sometimes gnaw off its own foot and escape; more important, humane societies dislike such traps because they imprison the rat underwater so it dies of drowning rather than being killed outright. Still, leg-hold traps are common for upland animals like foxes.

The week before setting his traps, Zehner, his friend, and I (J.S.) checked his trapping territory for signs of muskrat. By combing ditches and feeder creeks, he found rat runs and tracks and the small lodges muskrats build for shelter and giving birth to their half-dozen litters a year. Zehner then gauged how many muskrats lived in the territory and, therefore, how many he could expect to take without affecting next season's population. He estimated 800 muskrats in his 300-acre marshland of which he could safely take 500 to 600.

The weather was seasonable on December 1, lows in the twenties and highs in the forties, so we began laying the traplines. Muskrats stay in their lodges when the weather is very cold or when ice clogs the waterways. The lines along which traps are laid are simply muskrat runs, so that by following the line of the run or ditch we could find the traps. By law, trappers must submerge all their conibears at mean high-tide levels; submerging the traps makes it less likely that another animal, particularly a hunting dog, will get caught. Working at low and mid-tide, we carefully set sixty traps that day at fifteen- to fifty-foot intervals.

The next day we came back out to check traps and lay the last lines. Again by law, all traps must be checked daily, but trappers do so anyway because if one leaves a rat in a trap more than twenty-four hours the value of the pelt declines. Trappers have to stretch pelts on wooden frames as soon as possible or the pelt is almost worthless. That morning we picked up ten rats from the sixty traps and set the last forty. Zehner and his friend kept to the same schedule for the next six weeks and got almost 250 skins of which about 40 were black. It was a good early season, and it fit perfectly with Zehner's other activities—upland trapping and woodcutting, and his social life with the 4-H Club, the volunteer fire company, and his family in the evenings.

FIGURE 7. *Leg-hold and Conibear Traps*

A week before Christmas Zehner took twenty-five of his skins to sell for Christmas-present money to Harry Bakelee, a man to whom his father had also sold pelts and from whom Zehner bought some of his equipment. As Zehner drove in the driveway, Harry Bakelee emerged from his house with a fifty-pound bag of dog meal to feed the twenty beagles he kept in a series of pens near the outbuilding that served as his fur and equipment store. Bakelee, besides being a fur trader, was a rabbit hunter, fisherman, and gardener. He also raised beagles for field trials. He was a hale sixty-eight years old and was clearly glad to see Zehner, whom he had only greeted two or three times out on the bay during the past year. Zehner and Bakelee chatted briefly about weather, muskrats, and their families, while the dogs were fed.

The old man and the young man then went into Bakelee's rank-smelling shed to trade Zehner's furs. The shed held furs floor to ceiling. Skins hung everywhere; there were a pile of fur on the floor and a few unskinned animals in a heap. There were also a refrigerator filled with Schmidt's beer and two men sitting and drinking beer near the small wood stove. Pinup calendars—Petty Girls, Vargas Girls, Playboy Bunnies—littered the walls.

Bakelee has been selling fur for twenty-five years. He now sells to three dealers in Pennsylvania and two in New Jersey, and buys from about 150 trappers from all over South Jersey. In 1978 he bought almost 20,000 muskrat, 300 fox, and 100 raccoon skins. His largest buyer is an outfit in Hatboro, Pennsylvania, outside Philadelphia, that sells the backs of pelts for fur coats and the rest for felt. The black rats are especially valuable in Europe and Russia.

Zehner sold his 25 brown skins for $160, not a bad price, but he was right to hold on to the rest, since the price would go up. Then he headed home to prepare for Christmas.

SUBURBANIZATION

This was to be Don Zehner's last Christmas in Galloway Township. Caught between the suburban life his parents had chosen and the old life-style he prized, Zehner left for Maine the next year and has returned since then only for short family visits. As suburbanization continues, more young people will migrate to the three seats of yearning for the Pine Barrens—the state of Maine, the mountains of West Virginia, and the coastal plain of North Carolina—all of them rural sections where the pace and style of life reflects the longings of many young people in the Pines.

Bay people who live in coastal sections of the Pine Barrens have responded variously to land-use change. Some welcome suburbanization as an option, some mourn the loss of the past, the urban destruction of the barrier islands, and the seasonal crowding of the bays. Some create ways to continue the spirit of community life, and some leave. One key to the balance between new and old life-styles is the pace with which change comes. In hard-pressed Ocean County, residents have watched their environment deteriorate, and in reaction people like Janice Sherwood and Gladys Eayre of Waretown created the Pinelands Cultural Society, which holds regular Saturday Night Jamborees where people sing songs, tell stories, and trade recipes. Their band, "The Pine Coners," plays a regular schedule of appearances, and their intent is to create a sense of community. This is a difficult task in a place where old-timers are caught in an antagonistic relationship with outsiders who are called "webs"—weekender bastards—or "shoobies"—day-trippers who bring their lunch in shoe boxes. Joe King, of Bayville, on the central coast of Ocean County, put it this way:

> There was a time when a man could go in the bay and, in a day, get as many as 1,000 to 2,000 clams. Now he's lucky if he gets 400. It's tied to everything— everything has an impact on the environment. And these "weekend" clammers, they come down here, and they don't know enough to let a little clam grow up to be a big clam. They don't know, or they don't care, and they go out and buy a clam license, and they go out and rake a clam as big as your thumbnail, and then throw it in the basket, and they think they got a basket of clams. That's just a for instance.
>
> People come down here from the city and bring the city with 'em. They come down here, and right away they want to build shopping centers, bus lines, easy access to everything, and put wall-to-wall concrete in the Pines. I really have no idea why these people leave the city in the first place.

Joe is angered about the onslaught of suburbanization. Development is not new to him; he rejected it in favor of the old home place that retained the rural bay atmosphere. He is not against development—only its style, scale, and pace. Joe knows several old-timers who have built developments, and most, like Ed Hazelton of Manahawkin, meant real-estate development to be a small-time operation that boosted but not overwhelmed the local economy. Ed Hazelton saw home building as an adjunct to other activities, not as a way to destroy community life or change the landscape irrevocably. Three generations of Hazeltons and Cranmers (both families are original eighteenth-century settlers of the area) have built homes near Manahawkin and have participated actively in historical preservation.

Joe King and Ed Hazelton are in a dilemma that will become more familiar as this book unfolds. They resent sudden change while at the same time they defend the right of a person to do with land what he or she wishes. Bayside, along the causeway to Long Beach Island, was one of the area's first developments. C. H. Cranmer, Ed's maternal grandfather, built it in 1920, for both outsiders and local families. It was first called "Down Bayside," and Captain Tommy Cranmer had a clubhouse for duck hunters whom he guided on the marshes. The place was right next to Hilliards Landing, where farmers loaded salt hay onto freight cars. Cranmer built twenty houses that remain today. Ed said:

> The community has always been like one big happy family. Now the kids of the older people are livin' in there, and it's still quite pleasant. The older people like my dad and mother and others have passed on, but their children or grandchildren even have taken over. We still maintain it pretty much as a family settlement.

The floods and hurricanes that inundate Bayside from time to time have led to its nickname of "Mud City." Knowing the hazards of the sea, why would a group of old-timers own a place there? To Ed, it is a summer place, not a permanent home. He likes to think that he summers there just like the Indians used to. Bayside houses have been flooded out a half-dozen times in the past sixty years.

> Of course when you get water in your house, you can't do a thing about it; you just grin and bear it. You take everything out of the drawers and put it up on the table.

And then, when the tide recedes and the sun comes out, you hook the hose up to the water pump, and you just spray and take a broom and a mop and wash and dry everything all out. And start over again. That's all you can do.

There are a lot of cute stories about floods, kinda homespun things . . . where we had to get around the streets in sneakboxes and go from house to house and come all the way up to Hilliard Boulevard and come on up into town with your boat to get people's groceries and take 'em back down to 'em. A sneakbox is so low-slung, it's the perfect boat for that kind of thing.

I love the area, and I just want to do what I can to preserve it. I like to tell others about it.

The collective memory that links past, present, and future is strong along the coastal sections of the Barrens. Delaware Bay is distinctly different from the Atlantic coast. There are no barrier islands or protected bays, but only storm-battered marshes with poor access to metropolitan areas. There are few motels and campgrounds. This is not a tourist haunt, but an area of relic and unique indigenous water use. Oystering and salt-hay farming are traditional Jersey activities now found only along Delaware Bay, and in all of New Jersey, the sora rail, a small, black, delectable bird, is hunted only among the wild rice of one specific tributary.

OYSTERING

By the time Don Zehner was old enough to work the bays for shellfish, the oyster industry had almost died along the Atlantic coast. There was a time when oystering was important everywhere in brackish water, and the industry has had a long and vital history since the eighteenth century. In the early nineteenth century Yankee captains from Massachusetts sailed the Jersey coast and Delaware Bay dredging seed oysters to take back home where the oysterbeds had been overfished and exhausted. The New Englanders stimulated a thriving oyster industry along all the major estuaries, and in the twentieth century the coastal beds, never as productive as Delaware, which, in turn, was never as productive as the Chesapeake, began to disappear. What probably happened was that oystermen were taking both the oysters and their shells, selling the oysters to markets and the shells to the local iron industry as flux and to local farmers as a source of lime. Oysters cannot survive without dead oyster shells. More specifically, the free-swimming tiny oyster larvae that hatch in summer cannot "set" or attach themselves and grow unless they land on hard, gnarly substrate, namely, oyster shells. Where the substrate is silty, sandy, or gravelly, no set will occur, and the oyster bed will die. Long ago oystermen learned either to separate the live from the dead oysters and throw the shells back overboard or to barge shells of shucked oysters back out to the bays for setting grounds.

While the New England oyster trade was dying in the twentieth century, the Maurice River industry grew rapidly. By the 1920s, Port Norris, the industry's center, with its port at Bivalve, was a large community with two racetracks (photo 6). Oyster harvests in the 1920s were enormous—ten times what they were in 1980. In 1929, four million bushels were taken from seedbeds alone. The seedbeds in the region are north of the "southwest line," a point west of Port Norris that runs into the middle of the bay. North of the line the amount of sea salts in the water is 15 parts per thousand or less; farther down the bay, the amount of salts increases until it becomes 35 parts per

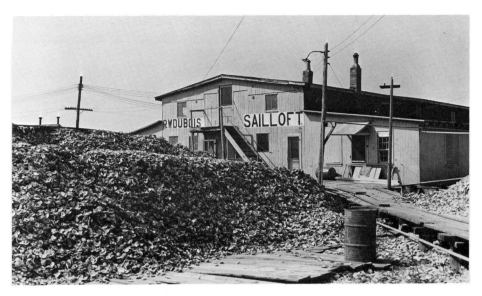

PHOTO 6.   *Oyster Shells and Sailloft, Bivalve 1934*

thousand in the ocean. Oysters grow most rapidly in water with 20 to 25 parts per thousand, but, unfortunately, so does a predator, the oyster drill, a snail, which can wipe out a seedling oyster colony by rasping through the thin shells of young oysters. When oysters grow larger, however, their shells thicken, and the drills cannot penetrate them. Since oyster drills cannot normally survive water less than 15 parts per thousand salt, seedling oysters are set out during "Bay Season" in May and June above the southwest line where, over the summer, they grow in size and thickness. In the fall, they are transplanted farther down the bay where they can grow quickly to eating size; leaving the small oysters above the southwest line, where they grow slowly, would make oystering economically unfeasible.

Oystermen take the summer off or go fishing while the breeding season continues from June to September, after which the main harvest occurs. There is a reduced harvest and a slow market from January until May when Bay Season resumes.

The 1920s were the heyday of oystering on the Maurice River. Then in the 1930s, drought, which struck the Midwest, also hit New Jersey. As the water level in Delaware Bay dropped, and as freshwater flows from the Delaware River slackened, sea-salt concentrations rose in the seedbeds. The oyster drills moved up the bay and began destroying the seedlings. Harvests dropped, and the Depression cut into market prices. Still the industry held on, even through a renewed drought period in the mid-1950s. By 1957 between 5 and 7 million pounds of oysters were harvested, and the industry looked at a rosy future.

In 1958 the oyster harvest plummeted 500 percent; the catch was 1.5 million pounds. The oyster beds had been invaded by MSX, an unknown parasite. Seedling transplants, in good health in the spring, began to turn into jelly, their cells dividing and proliferating, not like cancer clones, but gone haywire. Since no one knew the disease, it was called MSX, or multinucleated sphere, unknown. It was a parasitic protozoan, and though the disease began cropping up from Massachusetts to North Carolina, nowhere did it do the amount of damage that occurred in New Jersey and the Delmarva Peninsula.

The Rutgers Research Station at Bivalve has been working for twenty years on the problem. MSX now has a name, *Minchinia nelsoni*, and is known to decline at salinities less than 15 parts per thousand, but scientists still know little about its life-cycle, whether it has a vector, or why its virulence fluctuates.

The impact of MSX on the communities of the Maurice River was horrendous. Towns lost half their populations, and the boatbuilding industry became a shadow of its past. Port Norris is still an economically depressed area. Yet the communities on the Maurice hold on, their social structures intact, and the people are proud of their fishing heritage. Many in the oyster industry, both the white oysterers and the black shuckers, returned to the Chesapeake whence they came; some people dispersed to the cities, and others to local towns like Bridgeton, Millville, Vineland, and Woodbine. Many stayed and became proggers, the Chesapeake term for people who do different work during different seasons. Albert Reeves is a progger who has lived all his life among the people who worked the estuaries and the bay for oysters, built boats, farmed the marshes, and followed the seasons.

SORA

There is something special about the Maurice River and its major feeder stream, the Manumuskin. Partly, it is the enormity of the brackish and freshwater marshes compared to other tidal rivers, and partly it is the presence of extensive wild-rice stands (*Zizania aquatica*, locally called wild oats). The only other extensive stand of wild rice in the Pinelands is the Wading River basin, and that is now almost a relic.

Every evening, seventy-eight-year-old Albert Reeves told me (J.S.), he looks out from his back lawn in Mauricetown across the Maurice River, and every evening his wife asks him what he is doing, and every evening Reeves replies that he is looking at the river. What he sees in the river is his life.

Reeves was born in Mauricetown (pronounced Morristown or Mawrshtown) in 1902, and was moved to the house in which he now lives at the age of three, so he has been looking across the same back lawn for the past seventy-five years. The riverbed and its banks have not changed much since then, bulkheaded as they were by countless hulks of sunken and worn-out boats. When he was growing up, Reeves could see on the river large numbers of boats from the four-hundred-plus fleet that worked in and out of the Maurice, and when he looked onto the far bank he could see farms instead of cordgrass. The area's population then was at least double what it now is. Railroad stations existed at all small towns—Mauricetown, Port Elizabeth, Port Norris, Bivalve, Dorchester, Leesburg. The population fed itself on fish, shellfish, and especially the farms on the marshes. A 1919 map shows 50 percent of the marshes north of Mauricetown diked and drained to provide farmers the most productive soil in South Jersey, soil so rich in organic matter that it produced as much as an Iowa cornfield. But the dikes were difficult to maintain, and that type of wetland farming required a lot of labor. As the region went into economic decline with the arrival of the Depression and, finally, MSX, the last of the diked farms disappeared.

ANTHONY HILLMAN

FIGURE 8. *Sora Rail*

As the farms disappeared, wild rice replaced field crops, and what now exists on the Maurice River landscape is the wild rice of the slightly brackish and freshwater marshes, several old communities, and broad expanses of salt marsh frequently interrupted by major sand and gravel-mining operations. Of these landscapes, the most memorable are the wild-rice marshes and the towns.

The local wild rice is not collected for human use; it is a different variety than that which grows in the Great Lakes, and its small, tightly husked grains are not worth processing. These marshes are famous not for the rice, but for the sora, which feed on it during fall migration. The sora is a small, blackish railbird or ''mud hen'' with a delectable flavor (fig. 8). En route from their northern breeding grounds on inland marshes throughout the northeast, they stop only in the marshes of the Connecticut River, the Maurice, and a few other mid-Atlantic areas before wintering in the southeast.

For fifty years, between the ages of fifteen and sixty-five, Reeves ''pushed'' rail gunners in the marshes of the Maurice and Manumuskin in the same manner in which Thomas Eakins depicted the sport in his paintings of the area. Reeves was a guide for the famous and anonymous hunters who had heard that the Maurice held the best sora gunning on the East Coast. Between September 1 and November 1, Reeves earned a significant part of his income as a pusher, poling his double-ended gunning skiff through wild rice.

Sora, like all rail, are secretive and difficult to spot. They are at a gunner's mercy only at high tide when they have no firm footing. Reeves planned his early fall days around high tide. After meeting his customers at his house about two hours before dead high tide, he set out for a spot he knew would hold sora. Standing on one of the small raised platforms at either end of the skiff, he poled the boat upriver with one or two hunters armed with 20-gauge shotguns or 12-gauges with small number-one shot. Reeves would get to his spot an hour before high tide and would have that time plus about half an hour after dead high to hunt. Sora will flush only during that period; the minute the tide begins to ebb, the rails stay put and run through the grass rather than fly. Reeves was known as a great pusher because he knew where the birds were and how to retrieve them. Rail hunters do not use dogs because the marsh grasses rip them up badly and the footing is so poor the pusher must also retrieve.

Sora have a strangely weak flight. They pop straight up from their hiding place, flail the air in a frantic flight, and drop straight down like a stone to another hideout. The moment a gunner hit a bird, Reeves marked the spot with his eye, and, without taking his eye off the spot, poled and felt his way toward the fallen bird looking for clues, like the ragged top of a shot reed. Reeves had only two hours of high tide to help gunners bag their limit, which is now set at fifteen birds. When he first started pushing, the limit was fifty, and in his father's time hunters occasionally bagged more than a hundred birds in two hours when market gunning was allowed. Like other game birds, flocks of sora have declined drastically in the twentieth century, whether due to loss of breeding habitat, overhunting, loss of migratory habitat, or, more likely, a combination of all three.

Pushing may have been the most romantic occupation in Reeves's life, but it occupied only a small part. His chief job for almost twenty-five years was tender for the bridge between Mauricetown and a point on the east bank, just south of Port Elizabeth. Even during gunning season, both for rail and duck, he tended bridge and worked his pushing schedule around his tending hours. There were many times he worked for the town or county on a maintenance crew, and often he would help boatbuilders nail together boats or "cork" (caulk) them. Last, Reeves had done a lot of fishing, especially in winter and spring when he and a friend would drift the width of the river netting perch, catfish, striped bass, and shad.

FISHING

Like gunning, fishing is now primarily a recreational activity. Whereas another Maurice River man, Joe Stratton, used to fish commercially before 1950, he now makes his living with his charter boat that runs out of Port Norris. In fact, Stratton made a lot of money fishing and netting striped bass, or rockfish, served in local restaurants as "pan-fried rock." In the 1940s he had been able to buy his family a large house and several parcels of land in the area from his fishing proceeds. Rockfish are the most commercially valuable fish in the rivers and bays, but they run in cycles, and during the 1970s few have run at all in the area. They overwinter in the rivers and estuaries and, when spring comes, move in large numbers out the rivers into the ocean. The spring and fall migrations were the hottest fishing seasons on the coast, and fishermen could net hundreds of pounds during those seasons. Old men love to tell striper stories:

"Yeah, Old Man Henderson, Butch Henderson, my dad and I'd get out there with him. Remember when he used to net them stripers out there?"

"Yeah, he had a market for 'em. We didn't. He shipped 'em to Philadelphia."

"He was the stingiest man in the world."

"Yeah, so much a pound he got. I went by there one day and there was three of 'em haulin' a net, and I stopped and he said, 'You're just the guy I want to see. We need another man to help pull in the nets.' It takes four to haul, right? When you land the net, you've got to keep the net out tight and the bottom ahead of the top a little bit so the fish won't go out from underneath it. I said, 'All right, I'll help.' You're supposed to put a hundred pounds in a box. We caught him with, I think, eleven boxes of fish, and when I got ready to leave, he said, 'How many in your family?' I said, 'Four of us.' He gave me four perch about this long.''

''Yeah, his father, he got caught netting stripers. Paid about a thousand bucks fine. That was big money in the forties.''

Small groups of commercial fishing boats always did exist in small towns along the Jersey coast adjacent to the Pinelands region. Toms River, Barnegat Light, Atlantic City, and the Maurice River towns had larger fleets of twelve to twenty-five from time to time, but only Cape May had a fleet comparable to some of the small ones in New England and Carolina because only in Cape May is there enough deep water for large boats. The shallow inlets and treacherous shoals in most areas made it impossible to use the large boats commercial fleets need to catch their mainstay which are the enormous ocean and bay runs of cod, mackerel, blues, and weakfish, in addition to the deepwater lobsters and scallops.

So fishing remains a small cooperative operation. The fishermen and charter-boat owners keep each other abreast of fish movements, help each other keep the local markets in a healthy economic condition, and, when on the water, chat on their citizens-band radios.

Most coastal fishing is recreational. ''Summer ain't summer without bluefish, and winter ain't winter without a little perch.'' To local people fishing means one more meal without shopping at the market and, in particular, one more chance to be on the water, following the seasons. ''The stripers are running'' is synonymous with ''spring is coming.'' From the cold, early-spring days of perch, stripers, and winter flounder, the season moves quickly to the weakfish runs of late spring, and the flounder and bluefish of summer. In midsummer, when the waters of the bay warm to eighty degrees, the fish run out of the bays back into the ocean or up river into deep holes. September and October, however, are the premier fishing months as the blues and weaks move toward their winter grounds in Delaware Bay and off the Carolinas, the fluke return for a last forage and spawning in the bays, the stripers run back up the rivers, and croaker, taug, and sea bass are often feeding on bay bottoms.

By October's end, Stratton and other locals are tired of clamming and fishing and are ready for winter activities. They may take a foray up the rivers for perch, but such desultory trips are as much to check trapping territories as to catch fish.

To catch the largest perch, Stratton goes up the feeder creeks. Back in the maze of this morass he will occasionally stop his boat at a creek bend where he knows there is a deep hole. The hole was made by the current and tidal movement of creek waters gouging at the base of a bank stabilized by pilings. At mid- and low tide one can see pilings that someone a generation ago drove into the bottom. But why would someone drive piles in the middle of a salt marsh?

SALT HAY

Scattered throughout all the coastal marshes of New Jersey one finds evidence of such pilings, and a look at aerial photographs from 1940 shows what had been happening on those marshes. At one time people had built walkways and bridges through the salt marsh, but all that is now left are the old pillars that supported bridges and loading docks. Across this network walked farmers who had cut salt hay or spartina grass for a cash crop, and who had followed their fathers' pattern back to the early eighteenth century. A late-nineteenth-century New Jersey coastal atlas (Rose and Woolman 1878, 21) described the importance of what was then more than 155,000 acres of salt marsh.

> Their fertility seems to be inexhaustible as they produce annually, and without cultivation, large crops of natural grasses. These marshes are highly esteemed by the farmers whose lands border on them, as they not only constitute an unfailing source of hay for winter use, but also good natural pastures for cattle and sheep.
>
> Cattle keep in very good condition on them until about the last of June, when the green-headed fly, which is a native of the marsh, attacks them with so much severity as to prevent them from feeding properly. After these flies have disappeared, which occurs with the first cool northeast storm in August or September, the herds again improve in flesh and it is not uncommon to find them fat enough for beef in the months of October and November on these pastures alone.

No more than a half-dozen salt-hay farmers are left, all of them in Cumberland and Cape May counties on Delaware Bay. The rest of the farmers live only in old people's memories like that of Bill Wills of Green Bank and Farmer John Neal of Nesco:

> "Old Man Charlie Weber had a beautiful farm down there this side of Lower Bank. He lived off the farm and cuttin' salt hay. There's another industry that's gone to pot."
>
> "Yeah, over there, Thompson's Beach, Cape May, all them guys, the Coxes and Campbells, they always cut salt hay. Where I go fishin' they got a temporary roof built over it right on the meadows. And they got bales of salt hay, must be a thousand bales right there under that temporary shelter. And now accordin' to the paper, they're droppin' off. I don't know why."

The Coxes and Campbells still work the salt marshes on the Delaware, and the Berrys also work the Maurice River marshes. They keep the last of the salt-hay industry going, keeping ditches open and dredged so that tides can properly wash meadows, keeping foxtails (Phragmites) out and cordgrass healthy. Gas-driven machinery has, of course, replaced horse- or ox-drawn machines, and one can no longer see the old winches and bailers that used to sit out on the meadows; but after the July harvest, just before the spartina goes to flower, one can still see the long bales of salt hay standing in the meadow drying in the summer heat.

The Coxes, Campbells, and Berrys stay in the business because they like the work. Production in the meadows has decreased as the meadows themselves have become less productive, partly because firing them is prohibited by air-quality laws and partly because cheap labor, especially from family members, is no longer available.

The market for salt hay is now steady since it dropped precipitously after World War II when new industrial processes made cordgrass obsolete, and decreases in agricultural land meant farmers used less of it for mulch, especially on strawberries. Cordgrass mulch prevents strawberries from rotting in the fields, and, since seeds are not viable in fresh ground, it produces no weeds. Farmers and nurseries also use salt hay for hotbeds, and gardeners use it for landscaping. A few construction companies use it for septic fields and for preventing concrete from drying too fast or freezing, but only one man is left who still makes rope from it.

As Albert Reeves, then, looked out his backyard in Mauricetown across the marshes, he was not only looking across his life, but across a landscape, which once was more fruitful to past generations. Even the town in which he lives, like most along the coastal rivers, was richer in material possessions and social life than now. There was a time when the river towns were the main centers for coastal and Pine Barrens activities.

RIVERS, TOWNS, AND BOATBUILDING

When Captain J. R. Crowley was a boy in the 1870s, he once stood on the Green Bank Bridge and counted fifty-seven masts on the Mullica River, most of them on boats loading charcoal, wood, and glass for New York. Cap'n Crowley told the story to Rodney Koster who was born in 1907 in Herman City, cheek by jowl to Green Bank (photo 7). Over the course of Uncle Rod Koster's life, the masts disappeared, and the boat works, once scattered up and down all the coastal rivers, also disappeared as boatbuilding became the province of either single craftsmen or large factories.

Before 1920, before the automobilization of the United States, the most intense, varied, and rich life in the Pinelands lay along indistinct lines that connected the uplands of the Pine Barrens to the brackish and salty lowlands of the coasts—the river towns. Most still exist, but they now look more like outposts or suburbs than the regional centers they were 150 years ago. Toms River and Millville, both adjacent to the Pinelands Reserve, retain their original commercial and industrial eminence. What of Leesburg, Dorchester, Mauricetown, and Port Elizabeth on the Maurice River, Dennisville on Dennis Creek, or Somers Point, Bargaintown, English Creek, Tuck-ahoe, and Mays Landing in the Great Egg Harbor Basin? What happened to Pleasant Mills, Batsto, Crowleytown, Green Bank, Lower Bank, Harrisville, Wading River, Port Republic, New Gretna, and Tuckerton? Two things: First, the rural and industrial products of the Pines—wood, charcoal, iron, and glass—were either replaced by resources like anthracite, or could be more cheaply produced other places. Second, the automobile and internal-combustion engine made transport by boat obsolete.

When one looks back over the nineteenth-century census data and account books of Batsto and the Forks, where the confluence of the Atsion and Mullica rivers lies, an extraordinary picture of rich and busy lives emerges. At that time two famous Philadelphia Quaker families—the Richards and the Whartons—dominated the area. The Mullica Basin was a hub for rural industry, river transport, and boatbuilding.

Uncle Rod Koster remembers a half-dozen boatbuilders in and around Green Bank who built mostly garveys, skiffs, and sneakboxes. None now lives on the Mullica, and there are no more than a half-dozen boatbuilders along the Jersey coast and perhaps another six along Delaware Bay. Boatbuilders do other seasonal work, but local people know them for their boats, which are bought by people all over the United States. With the renewed interest in crafts and folkways since the 1960s, some builders have been besieged by young people wanting to learn their techniques, so the old traditions are unlikely to die. In the past decade a spate of books and articles on boatbuilding and decoy carving has appeared, and some craftspeople complain that, since they spend so much time telling people how they do what they do, they have no more time to do it. In

PHOTO 7.   *Rodney Koster at the Herman City Hotel*

©*James F. Gandy, Jr.*

the past a wide variety of designs for boats existed from the exotic-sounding piraguas, shallops, and pinnaces to the rustic catboats, scows, and melon seeds. Small-timers now build only garveys, sneakboxes, and sea-bright skiffs.

In the nineteenth century the Richards's boatyard at the Forks near Batsto produced schooners well over 100 tons, although most were in the 30- to 60-ton range. So many were produced in South Jersey that in 1900 New Jersey had the largest working sailing fleet in the United States. In the Maurice River old and unsalvageable boats were sunk along the riverbanks to stabilize them, so that these hulks literally line the river. Old hulls and ballast litter the muddy bottoms of all the rivers.

Boatyards needed a wide variety of materials, some in huge quantities: oak, cedar, and pine for planking, ribs, decks, and masts; teak and mahogany for the railing, paneling, and cabins; cordage, iron, and cotton for the sails, rigging, fittings, and anchors; glass for the windows; tar and turpentine along with tallow and sulphur for the caulking and for bottom paint. Boats also needed provisions and furniture, which drew on local farm produce and on local cabinetmakers and coopers. Many other skills were required: shipwrights and carpenters; sawyers, joiners, and carvers; caulkers and sealers; instrument makers, glaziers, and blacksmiths; and chairmakers and upholsterers. Many people could, of course, combine talents, and few boatyard workers depended solely on boatbuilding for their livelihood. The market for boats, as for all large capital-intensive products, is at times uncertain.

Likewise, today boatbuilding in the river towns is uncertain, but companies remain in the area because in few other places can they find a combination of proximity to water, a central location for eastern markets, and a work force with generations of craftmanship and technique to support it. Pacemaker, for example, the largest boatbuilder in the Mullica Basin, is in Egg Harbor City and hires two hundred to five hundred people. The hulls of their motor yachts, between thirty-three and sixty-six feet, are fiberglass, but everything else is wood, and most of the labor force is occupied with engines and woodwork. They are proud that one could never find a gap in the joints or a flaw in the teak work.

Uncle Rod's favorite place is the Forks, where the old boatyard used to be and where he can go in quiet and loneliness. The Forks is Mullica River's Janus. One face looks downriver, where eighty-foot schooners were launched toward the sea; the other face looks up into the freshwater swamps of cedar, which grade into the pitch-pine lowlands and upland oaks. Uncle Rod has lived his whole life in this transition zone, half in saltwater, half in fresh. He is half bayman and half Piney, half clam shucker and half stump jumper. He has chosen to live out his life where he was born, in the Herman City Hotel that his grandfather, John Rapp, helped build in 1869 and in which his father, Leon, was born. The hotel is a magnificent, rambling, two-story structure, one of the few saved from frequent fires that swept through the Pines. Uncle Rod and his various dogs occupy three downstairs rooms of the place. The hotel was built in conjunction with a glasshouse, or factory, that Charles Hermann, secretary-treasurer of the German Society in Egg Harbor City, began in about 1870. The glasshouse was completed in spring 1873. That summer saw the 1873 Crash, the worst depression to that point in history, and the Herman City Glass Works went bankrupt in the fall. The venture barely lasted six months, but it left the Herman City Hotel as a legacy.

It was in 1873 that Joseph Wharton began buying land in the Mullica watershed, for what reasons we shall soon see. Uncle Rod's father found himself working for Wharton, doing odd jobs around the mansion and working the cedar forests. Uncle Rod was brought up with the same intimate knowledge of the area his father had passed down to him. When New Jersey bought the Wharton estate's holdings in 1950, the state turned it into a state forest, and Uncle Rod became one of the first rangers. He retired from work in 1977. Father and son had not moved from their birthplace. They had lived out their lives, one foot in the Pine Barrens, the other in the estuary.

Don Zehner, Ted Von Bosse, Albert Reeves, Rodney Koster, and Ed Hazelton are of different ages and backgrounds. They come from geographical areas so separate that they know neither each other nor each other's hometowns. All of them, however, share the rich coastal and river resources; they take part in the seasonal cycles and collective memory of the coastal sections of the Pine Barrens and know something of the Pine woods as well as the bays. Just as these men would lose much of their life-style if the coast and rivers of the Pines were degraded, so the Pines would be diminished without them and their colleagues.

Freshwater: Regional Control

BOG-IRON MINING

In April 1781, according to the Batsto Furnace logbooks (Pierce 1964), William Richards, ironmaster, was on his way by stage from Batsto to Philadelphia. It was a two-day journey, and Richards broke his travel at an inn on the Jersey side of the Delaware River. He had left his manager in charge of the furnace while he was in Philadelphia on family and commercial matters. As ironmaster and owner, he spent much of his time at the mansion in Batsto, but still regarded Philadelphia as home.

Richards's mansion was actually a rambling, two-story wooden structure that stood hard by the furnace itself, and from his porch he could see almost all his operations, for the mansion was the seat of authority and center of social life for the village of some five hundred people. The mansion housed his own family plus his servants, and the basement served as a living and dining area for unmarried workers. Nearby were a gristmill and a sawmill, smokehouses, a spring house, barns, bake ovens, a garden and orchard, and a summer house. Across the millpond, down from the sawmill, were the workers' cottages, gardens, and outbuildings.

Batsto was the seat of industry for the Richardses who ran a paternalistic empire from that village. Iron-furnace towns were, in fact, called "plantations" with all the connotations of that word. Workers came from other forges in the region, from Europe, and from the children and relatives of current or former employees. Long tenure of skilled workmen was common. Life in a forge village offered plenty of work, ample pay, reasonable conditions, and some leisure time with an active social life (Walker 1966).

Richards's journey had taken him through lands his family owned and would acquire over the next two generations. At their peak, the properties of the Richards family totaled more than a quarter of a million acres. Through exploitation and investment in Pine Barrens water and woodland resources, the Richardses greatly influenced industrial development in the Delaware Valley. They built ironworks, glasshouses, cotton mills, brickyards, and boatyards. They made munitions for two American wars and supported churches, schools, the University of Pennsylvania, and scientific societies. They helped establish the Camden Atlantic Railroad, which created Atlantic City, and organized two of Philadelphia's largest banks, some of the major insurance companies, and other financial institutions. Family members served in the New Jersey and Pennsylvania legislatures, and one was mayor of Philadelphia.

Batsto, along with the rest of South Jersey and eastern Pennsylvania, was an integral part of a larger commercial system. From the first days of William Penn's colony, a landed, wealthy, highly educated group organized the use of the surrounding regions into that system. A single-minded devotion to commerce, exceptional business acumen, and an early start combined to give the Philadelphia Quakers a large share of America's early commerce (Tolles 1948). After London and Bristol, Philadelphia was the busiest port in the worldwide British mercantile system. Quaker merchants developed triangular and polygonal trades, shipping Philadelphia's products to countries all over the world in exchange for other goods. A ship built at a boatyard in the Pinelands might be sold in Europe or used to transport cargo worldwide.

Trade flourished, and so did the merchants. Bonds, mortgages, real estate, agriculture, forestry, and mining offered the best investments for surplus capital, and from the 1720s onward, the iron industry gradually overshadowed all other forms of industrial investment. The largest share of capital that supported iron production came from Quakers who became ironmasters for many of the furnaces and forges.

William Richards learned the iron trade at a Quaker forge near French Creek, Chester County, west of Philadelphia. His first investment was Batsto which he purchased from a West Jersey Quaker, Charles Read. In Pennsylvania's Piedmont Richards had learned about iron ores formed by igneous and metamorphic processes, but in South Jersey's Pine Barrens he would work with limonite, or bog iron, which forms in water.

Bog iron comes in several forms that grade one into another. Loam is reddish flecks of iron in sand; seed ore comes in partially consolidated layers; and massive ore is a rocklike concretion found under swampy deposits. Iron furnaces used only the latter two which were gathered by ore raisers in long, shallow-draft, double-ended barges. The raisers first located a promising bog (photo 1), often in an area called a "spung," or a spongy swamp that drains internally instead of out to a stream. The raisers then diked a section, drained it, and with poles and brute strength, "raised" the massive and seed ore out of the bogs, into the barges, and back downriver to Batsto.

Besides ore, furnaces needed limestone for flux to pick up most of the impurities in the ore and charcoal as a hot fuel for processing. Limestone quarries, coastal Indian middens, and oystermen provided adequate flux, and the surrounding forests provided wood for charcoal. Until the advent of anthracite, wood was the only available fuel, and the amount of wood available in any one area limited iron production. Although limonite is inferior to the magnetic ores of the Piedmont, the Pines had extensive woodlands, while farmers in the Piedmont had cleared most of the forests. Until the appearance of railroads and coal in the 1840s, some of the largest and most important ironworks in America were in the Pine Barrens where the concurrence of bog iron, pine woods, abundant water, world and regional demand for the resource, and Quaker investment produced a rural industrial empire.

To operate their furnaces and forges, the Richardses needed an average of about 8,000 acres of woodland per year for each complex the size of Batsto (Sinton 1977). This accounts for the need to have 75,000 to 100,000 acres of woodland surrounding each site and for the outside control of so much land in the Pine Barrens even today. The Pine Barrens were to a great extent a mercantile colony of Philadelphia, and from the early 1700s people outside the Barrens would decide much of the area's future.

The iron industry prospered from 1760 to the 1840s when new fuels and technologies ended its production and America's industrial base began to shift from rural areas to the cities of the eastern seaboard. The 1873 depression wiped out most of the remaining rural South Jersey forest- and water-based industries, and large tracts went on the auction block at very low prices.

PUBLIC WATER SUPPLY

The major buyer for this unwanted land was a fourth-generation Philadelphia Quaker, Joseph Wharton (1826–1909), who purchased more than 100,000 acres formerly owned by the Richardses, including Batsto and Atsion, over a twenty-five year period. Wharton was part of the American elite that led the Industrial Revolution and provided a source of leadership and counsel to the nation. The Wharton family retained their Quaker faith and aristocratic interests in profit, investment, and public service as witnessed by his founding of the Wharton School at the University of Pennsylvania. Wharton's ventures straddled the era of industrial partnerships and the beginning of modern corporate America, and his holdings in the Pine Barrens continued the tradition of outside control for a major part of that region.

Wharton's land purchases in the Pines were no idle or foolish speculation. In the 1870s Philadelphia faced a major water-supply crisis. A blue-ribbon panel of experts (Philadelphia 1875) appointed by the mayor and funded by the state legislature

> estimated the probable amount of impurity likely to be thrown into the Schuylkill [Philadelphia's potable water source] and [we] quote the arguments and evidence which, in our opinion, establish the position that once fouled to the degree we have supposed, the water can no longer be properly recommended as a drinking water for the city. (2)

Philadelphia, like other eastern cities, had polluted its own drinking water. Wharton was well informed about the water-supply crisis. The reports were published public documents, prepared by a scientific community whom Wharton supported. He may well have known the actual scientific investigators, and he also knew that his own industries upriver from Philadelphia contributed to the river's pollution. Two years before the mayor's office issued its report on June 5, 1875, Wharton had begun buying Pine Barrens land at bargain prices, land that covered a large part of the Cohansey aquifer and much of the surface water of South Jersey's largest river, the Mullica.

The 1875 report recommended using fresh mountain water rather than treated river water, and offered detailed maps and cost studies of aqueduct systems to bring it from as far as the Pocono Mountains, yet Wharton's scheme for the Pine Barrens offered a cheaper and closer supply than Pocono water. Water-resources technology of the period, the amount and quality of Pinelands water, the intensity of the crisis, and Wharton's investments made the Pine Barrens scheme quite feasible. Only a last-minute action by the New Jersey legislature foiled Wharton's plan to export water from the Pine Barrens (map 5).

In 1884 the New Jersey legislature unanimously passed a water-resources act, section 21 of which made it unlawful to

> convey any [state] waters beyond the limits of this state or sell or dispose of any of the rights, privileges, or franchises to any person or persons or corporations for such purpose.

The law actually substituted one outside control for another. If Pennsylvania could not have the water, perhaps North Jersey could? Outside speculation on land in the Pine

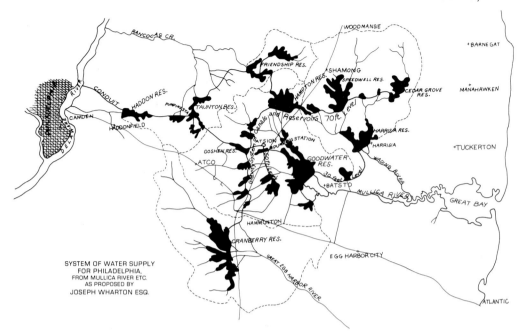

SYSTEM OF WATER SUPPLY
FOR PHILADELPHIA,
FROM MULLICA RIVER ETC.
AS PROPOSED BY
JOSEPH WHARTON ESQ.

JOSEPH WHARTON'S WATER SUPPLY PLAN FOR PHILADELPHIA

MAP 5

Barrens continued, but Wharton's heirs soon tired of different South Jersey projects, and in 1915 offered the 100,000-acre tract to the state for $1 million. State officials called a referendum on the purchase, and voters defeated the question 123,995 to 103,456 with almost all South Jersey voting against the state's purchase. In 1950 New Jersey did purchase the tract, almost 2 percent of the state's total land area, for over $3 million (Pierce 1964).

The most obvious reason for South Jersey's negative vote was the loss of local tax revenue because the Wharton family paid considerable taxes, whereas the state only paid ten cents an acre per year to local towns for the Wharton-tract land. Wharton had made himself popular with many local residents. He was personally courteous, his projects provided local employment, and he allowed local use of his extensive lands for hunting, trapping, gathering, and fishing. Further, his ownership continued the century-long pattern of plantation ownership by prestigious Philadelphia families and well-known local managers. Public ownership promised none of the benefits of Wharton's ownership and all of the drawbacks. Local residents believed in gainful employment based on local resources; this work was their life. Public ownership meant an end to this pattern and another form of outside control. It was never clear that public ownership served the best interests of local people, and it still is not.

Since the turn of the century, ownership continued to follow a pattern of outside control, and new residential and commercial developments were valued more highly than resource-based activities. The twenties marked a modest decline in forest industries and an increase in land frauds based on the sale of the extensive old industrial tracts. The result was thousands of tiny scattered plots on tax maps, unclear titles, defrauded owners, and tax-delinquent sales. A significant proportion of landowners still live outside the region.

Since World War II, the Pine Barrens have seen schemes for international jetports, new towns, state and national parks, the Air Force Academy, amusement theme parks, water-diversion plans, and preservation plans. Each in some way had called for the diversion and manipulation of the ground and surface water systems. Preservation called for artificial wilderness regulations, new towns required impoundments, prestigious lakeside developments, potable water supplies, and sewage disposal. The jetport was to have been sited, along with a new city and large state park, on the area of greatest interbasin transfer. Every scheme involved outside ownership and state participation in environmental manipulation. In many cases state investment, coupled with local zoning regulations and speculative ownership, had already resulted in polluted waters and destructive residential growth. With every ''scam,'' as the locals called land speculation, it was never clear what was the role of the state and the private investor. Such continued manipulation, coupled with inherent distrust of government, led to pervasive regional cynicism about the control and conservation of land in the Pines.

The 1978–80 federal, state, and local effort to create a national reserve has run head on into the century of cynicism. The land-management scheme proposed in the Pinelands Protection Act of 1979 emphasized protection and enhancement of the significant values of Pinelands resources, but the definition of significant values comes chiefly from outsiders even though the act required maximum feasible local-government and public participation. Local residents feel that any land-use plan must address the questions of financial equity, government responsiveness, and local participation in resource decisions as well as preservation of natural resources.

Wharton's plans and the actions of his heirs left the people of New Jersey their largest tract of state-owned land and preserved a large portion of the central Pine Barrens intact. His real legacy to South Jersey, however, may be the current internal agenda of the local residents that has its historic roots in the financial manipulation of the Pinelands by Philadelphia merchants and the state of New Jersey. It is no accident that locals cynically see some leaders of the preservationist Coalition to Save the Pinelands in the same light as they see the state and the speculators.

Had Wharton's plans succeeded, the central Pines would have become a series of large lakes and reservoirs linked by rivers and canals. As a water-supply area, public access would have been restricted. Instead, his plans inadvertently led to the creation of Wharton State Forest. Since the 1950s the forest has protected the rivers which Wharton would have dammed and diverted. The free-flowing little rivers have become the basis of a substantial recreational economy and have remained the foundation of heavy outside use and participation in the life of the Pines. Through recreational use of the rivers, longtime Pinelands families come in contact with people from all over the world and especially the mid-Atlantic region. One of these is one of the authors of this book, Jon Berger.

Insiders and Outsiders, Part 1

RECREATIONAL CANOEING

I (J. B.) am a canoeist. Since the age of twelve I have canoed the rivers of northern Canada to James Bay and Hudson Bay and the streams and rivers of New England. My first acquaintance with the Pine Barrens came on a day I canoed the Wading River.

©*James F. Gandy, Jr.*

PHOTO 8. *Bass River*

Canoeing Pinelands rivers is a unique experience that provides relaxation, good technique, and gradual conditioning for summer cruises. The varied vegetation of the rivers and swamps, the white gravel of the shores, the black mucks, and the clear amber water are seductive.

Rivers in the Pine Barrens are classed as "easy and not challenging, suited for novices" (ARCC 1981). However, I have always been leery of river-classification systems because they seem to be based on the difficulty of shooting rapids and not on the skill necessary to negotiate any stretch of river under particular water conditions. As a wilderness cruiser I probably portage or line and wade down many rapids that others might shoot, but I also know that small streams with good current and meanders, varied depths, and overhanging vegetation are tricky and interesting. Threading one's way through sandbars and shallow water into a headwind is far more difficult than running a large rapid with big waves and strong current. The rivers of the Pines provide the expert paddler a sophisticated day of maneuvering (photo 8).

Rocks are no problem, but downed trees and stumps are. I pride myself on being able to negotiate the narrow streams and their swift currents, hardly touching submerged logs or live branches that lean out from the banks. This requires careful reading of the water and coordination of the stern-paddler's stroke as the canoe enters river bends. I follow a center line down the stream and do not let myself get too far down into the turn, where the current can wash the canoe into the bank and overhangs. Because I advance relatively slowly and keep room on the downstream side of the turn, I can make allowances for submerged logs at the turn of the bend. When I see a clear passage, I can give a burst with the paddle and flow around the bend and onto the next one. The whole maneuver requires the coordination of paddling speed, the river's flow, the shape of the channel, the density of the vegetation, and the anticipation of recurrent bottom, bank, and flow patterns at the turns.

I like going to the Barrens. My wife and I can even canoe in the low water in the fall, because Pinelands streams receive plenty of groundwater to maintain their flows. The fall is even better than the spring. In spring I get the feeling of renewal, but in fall I enjoy the changing colors of blueberry bushes and the feeling that we are getting in one last run before winter.

I usually rent canoes from George Mick, of Jenkins, who has one of the eighteen canoe-rental businesses in the Pines (photo 9). Mick is in his fifties and lives at Jenkins, in Washington Township, Burlington County. He was born in the house next to the canoe-rental agency and gas station/store. His ancestors were Irish laborers who came to the Pines in the late-eighteenth or early-nineteenth century from Ireland by ship and quite possibly landed at Tuckerton, which was then an official port of entry along with New York and Philadelphia.

From Tuckerton the Micks moved to the nearby iron-furnace villages and worked for the Richardses at Atsion, Hampton, and Martha. The Martha Furnace diary from the first decade of the nineteenth century mentions the Mick family. Martha was

PHOTO 9. *Mick's Canoe Rental, Jenkins*   ©*James F. Gandy, Jr.*

located on the Oswego River, also called the East Branch of the Wading, a tributary to the Mullica, and Mick has said that his family "has been here forever." On the Oswego is a spot called Mick Place.

Today the George Mick family runs the canoe-rental agency full-time, and they drop people off to canoe the Oswego and Wading Rivers. In almost two hundred years they have not moved far from where they settled; the store and livery have been fifty years on the same spot, but the family only began to rent canoes in 1972. Before that they owned and cultivated blueberry fields while the store at Jenkins catered to the small number of locals and to the many hunting clubs, but canoe renting proved to be a more amenable way to make a living than blueberry cultivation.

In 1972 Mick decided that on hot, sunny days he would rather drive people and canoes around than pick berries. With his brother Howard and their wives they began the rental business, which, like any other line of work, has ups and downs. In the Micks's business "you have to make people happy." Paradoxically, the timelessness and slow pace of canoeing does not permeate the operation or the perceptions of many people who come from all over to canoe. Everyone wants to get out on the water as fast as possible and be picked up "just when they want." When things get crowded on summer and fall weekends, people fume and curse, waiting in the parking lot as others load up and leave. People want "promises about when they can get on the water and what the water will be like." The Micks have over 250 canoes, and on some weekends all the canoes are rented. On those days work begins at five in the morning and ends at ten at night.

Mick estimates that between his agency and another downriver run by the Bell brothers, they must take out over 5,000 trips a year. Each trip may average about three canoes; that means that over 15,000 separate canoes go down the Batsto, Wading, and Oswego rivers from March through November from two agencies alone. There may be between 30,000 and 45,000 people involved in the trips. Groups come from all over the East Coast, from as far away as Connecticut and Maryland and a few from the rest of the country and from Europe. Many are from organized clubs, churches, schools, and scout groups, but the majority consist of families or informal groups of friends who come year after year.

There are also people who represent powerful outside interests, such as members of the Sierra Club, Friends of the Earth, the Environmental Defense Fund, and the Audubon Society. Canoeing is often an outsider's introduction to the Pines, and once an outsider's interest is piqued, he or she may return often for recreation or scientific study. More important, the outsider may take an interest in the politics of Pinelands preservation and acquisition of land for public use.

Almost all the Wading and Oswego rivers lie within Wharton State Forest, and most canoe travel occurs on these public lands, with some exceptions when the rivers flow past privately owned cranberry watersheds. Thus the condition of roads and maintenance of river corridors are thought to be the state's responsibility. From earliest times sand roads connected the rivers and followed their courses. Early travelers, woodcutters, and colliers established the honeycomb pattern, and some roads were stage routes between Philadelphia and the coast. Now the state owns the unpaved roads, and with hard use from four-wheel-drive vehicles, canoe hauling, and the weather, these roads become filled with deep potholes. Sometimes the Micks find it impossible to get their clients and canoes over the roads. The Micks believe the state should maintain the roads, but state personnel reply that the condition of the roads restricts overuse of the area.

Trespassing, solid-waste disposal, and seasonal overuse are other issues. Where the rivers pass private cranberry land, there is always a chance of abuse, and the Micks hear most of the complaints from the cranberry growers. Beer and canoes often go together. Through the clear, amber water one can see cans on the bottom, in some parts forming aluminum veins down the river. On those days when all 250 of Mick's canoes are on the water, it looks like there are aluminum veins on the water's surface. George Mick expressed the fear that increased traffic might cause the state to limit access or to charge for permits during peak times. These problems are not unique to Pine Barrens rivers, but added to the work of keeping people happy, they make the days between March and November so busy for the Micks that the December influx of deer hunters seems like a holiday.

Almost all eighteen canoe liveries in the Pine Barrens are family run, and all specialize in particular rivers, watersheds, and road networks. The distinctive juxtaposition of wet and dry ground in the Pines meant that early road networks focused on the watershed unit. Therefore, many sets of roads follow the dry rims just off riverbanks or the higher ground at the borders of watersheds. The perfect canoe landscape is a series of bridges and interconnected roads that afford a number of drop-off and pickup points for all levels of canoeing interest. The short haul and sinuousness of the rivers mean that in the best of circumstances a half-hour drive or less will allow canoeists to start on an eight-or ten-hour trip that may even end right at the canoe-agency dock.

The eighteen rental agencies represent a large and growing national interest in outdoor recreation. All of them, except one, were started after 1970. Hacks of Mount Holly began to rent canoes and rowboats in 1875. Hacks sits on the Rancocas, just on the edge of the Pinelands and closely connected to Philadelphia by rail and highway. The current proprietors say that the Rancocas is the ''aristocrat of canoeing. There are no logs to pull over, no little bends to negotiate, just a placid stream, controlled by dams, that flows through privately owned land, well kept.''

Since the nineteenth century, well-to-do Philadelphians have taken summer retreats to country homes located near water. For some this meant the Jersey shore; for others, the nearby lakes of the adjacent Jersey inner-coastal plain. This entire series of impounded waters, used by early industrialists, later became used for recreation. At first rowboats, lace collars, and portable victrolas were the norm for a genteel day on the millpond. Canoes then became popular, and the rise of scouting began to bring people to these areas to satisfy a bit of wanderlust. By the late nineteenth century Americans had begun to believe that the out-of-doors was not a fearful, but an enjoyable place. Papa Hack at Mt. Holly began to haul canoeists first by wagon and then by car and truck to the headwaters of the Wading and Mullica rivers. Scout troops would spend the weekend canoeing down to the coast, and Hack would meet them on a Sunday to bring them back. During the twenties and thirties Hack had over 140 wood and canvas canoes and long waiting lists each weekend.

In 1942 Margaret and James Cawley, now in their nineties, published their first edition of *Exploring the Little Rivers of New Jersey*. The Cawleys represent a group of people who have spent much of their leisure time on the rivers of the Pine Barrens; they are the first American generation who took part in the Progressive Conservation movement of World War I and the New Ecology of the 1960s. Their mission was to expose people to the benefits of the environmental movements through activity and education. The Cawleys provided part of the stimulus and motivation for people to go to Hacks and later to Micks to enjoy the rivers of the Pines, and they certainly contributed to the almost insatiable American demand for nonmotorized, clean, water-based recreation.

People involved in outdoor recreation have an interest in clean water. Outsiders rely on the clean rivers a few days a year, hoping they will always remain free flowing and clean. Coming for a day's canoe trip on the Oswego, they see the Pine Barrens as "pristine"—uninhabited, wild, and clean except for litter. They want the Pines to remain exactly the same, but as a Piney said: "If you don't want 'em to change, better take a photograph, for they'll be different tomorrow." The insiders have a far different perspective.

Outsiders are not the only ones who use the waters that flow in and out of the rivers. There are also residential and commercial users downstream in hamlets and towns, and there are the agricultural users. "Clean" water is a relative concept, and "pristine" is a loaded term that scientists, lawyers, and planners on the Pinelands Commission have translated into a precise standard, namely, water with very low (.17 ppm) nitrogen concentrations suggesting little or no human activity in the area.

The major users of water in the Mullica/Batsto/Wading watersheds are cranberry growers, who need clean, but not pristine, water for their bogs. They must use pesticides and herbicides, so they contaminate the pristine quality of water flowing out of the bogs as does any farming enterprise. The water from cranberry bogs is still clean, meaning potable and capable of sustaining the unique habitats of the Pines, but cranberry growers have no use for outsiders' concepts of a pristine, uninhabited, and wild landscape. Still, they know better than anyone else that clean water requires hard work and vigilance. Among the most vigilant is Mary-Ann Thompson.

The Heart and Conscience of the Community

CRANBERRY AGRICULTURE

On a cool, clear morning in October the mist still lies over the Birches Cranberry Company bogs even as the sun illuminates the scene. Soon the rising sun will burn off the mist and reveal a maze of connected reservoirs, bogs, canals, dikes, spillways, and dams (photo 10). Only the dikes show up as slight rises above the mist, and the reservoirs and other open water are marked by the distinctive pattern of shiny light. Mary-Ann Thompson looks out over the scene with pride and relief. Perhaps her efforts and those of her neighbors have come to fruition, and they have convinced the state to protect these bogs and contiguous watersheds and forests from the indiscriminate development that has occurred adjacent to the Birches. For cranberry growers, water, its physical and biological qualities, its aesthetic and historic memories, its economic value, and its deep emotional meaning, are of prime importance.

As the morning wanes, the cranberry harvest will continue. George, a Puerto Rican, will operate the forklift. Leon, a Piney who lives on the bogs, and Gladys, a black woman who drives a labor bus from Philadelphia, sort the berries by hand.

PHOTO 10. *Aerial View of Hog Wallow Cranberry Bogs*

Wilbur, Jr., a Piney and the mechanic, tends the machinery. Lennie, Wilbur, Sr., Butchie, Jimmie, and Pudgie—all Pineys who live throughout the Thompson watersheds—drive trucks, work the harvesters, and collect the berries. Different members of the Thompson family check water levels, oversee the operations, and lend a hand wherever needed. A small cannery in Hammonton waits for the berries because this is one of the few cranberry operations that is not part of the giant Ocean Spray cooperative. All these people depend on the water and its management for their livelihood, and their emotional and economic lives are tied to the future of the water supply. Protection of the water resource has led to an integrated system of agriculture and to major political splits between growers over the use and regulation of the Pine Barrens.

A barren, in agricultural terms, is any land that cannot produce crops. Early Pinelands settlers found the sandy, acidic, dry soils of the uplands and the water-logged soils of the lowlands low in productivity compared to adjacent soils on the inner coastal plain. On lowland soils in the Pines only native cranberries and blueberries have prospered on a large scale, even though the Pines lie at the southern limit of their natural distribution (Applegate et al. 1979).

The wild cranberry, *Vaccinium macrocarpon*, a wild trailing evergreen vine, is native to sandy, peaty bogs of North America. Early agriculturalists gathered them for domestic consumption and commercial use. In 1835, members of the Thompson family planted the first cultivated cranberry bog in the Pine Barrens, and in the following hundred-year period, growers cut down or flooded out thousands of acres of white-cedar and swamp hardwoods, and converted almost every first-order stream to productive bogs. There was a cranberry craze in the mid-nineteenth century, when anyone with a piece of wetland would install a cheap wooden dam and grow cranberries behind it. As happened so often in the free markets of the period, the result was a glut, and cranberry growing was prone to the same wild fluctuations as were other agricultural products. Most of the small-time cranberry operators were out of business by 1940. Cutting and ditching drastically reduced the amount of white cedar that once grew in the Pines. In their place are millponds or lakes that provide fishing and boating, or, more often, overgrown freshwater wetlands that provide habitats for orchids, curly-grass fern, and Pine Barrens tree frogs. Of the unknown hundreds of cranberry growers who worked the wetlands two generations ago, there are now only fifty-five who together farm more than 3,500 acres of bog land.

Cranberries need a regulated water supply and, unlike field crops, require year-round maintenance. The shallow-rooted vines are susceptible to frost, so growers use water-control systems to avoid seasonal frost damage. The water they use to cover the vines throughout the coldest winter months is called "winter flood." Spring signals the "spring draw," or removal of water. The degree of predicted frost on any given night determines the proper water levels necessary to protect new buds, and some growers have installed expensive sprinkler systems. Temperatures on cranberry bogs are often ten degrees colder than on the surrounding uplands, and frosts in May are not uncommon.

As the growing season progresses, the flowers begin to develop. First the buds become "dangles," which open to blooms that progress to "fruit set" followed by green, yellow, and red berry stages to maturity. During the season, drought can lower water tables and require growers to use extensive irrigation systems that tap streams or wells to protect the soil from drying out. Growers need a guaranteed regional water supply so they can switch to pumping groundwater should surface flow become insufficient at the height of the growing season.

While cranberries need to have their roots constantly wet, they cannot survive the damaging floods sometimes brought by spring and summer thunderstorms. The system of dams and ditches serves as a flood-drainage and control system. Summer flood hazard eventually gives way to fall frost problems that can begin as early as August. A combination of summer drought and fall frost can prove disastrous, and growers may not have sufficient water on hand to combat the frost. Again regional water supply is necessary to transport water in from distant parts of the watershed.

As berries ripen, their ability to survive lowered temperatures increases, but below twenty-five degrees lack of water protection can mean loss of an entire crop. Toward the end of September, the harvest can begin as temperatures decline rapidly. Growers often plant later-ripening varieties closer to the reservoirs, where the warmer water will protect the latest berries.

Fall flood is essential to harvesting the berries, which are picked "wet." Before the 1950s, cranberries were "dry picked" by a large labor force walking across the bogs with toothed cranberry scoops. Now growers flood the bogs and use mechanical harvesters to knock the berries from the vines. Water barely covers the vines; the berries are knocked off and float on the surface where they are gathered into a small area by the use of long boards and removed by mechanical scoops into trucks. Each bog in turn receives its supply of water as one after another is picked. The speed of the harvest depends on the amount of water available; an insufficient supply can cause a delay.

After harvest the bogs are readied for winter flood, and the cycle continues. The cycle with its attendant technologies, participants, and products has led to a distinctive cranberry landscape. The major components are watershed lands to protect the bogs, the intricate connections of bogs and hydraulic works, and the cranberry village.

The cranberry landscape is, above all, extensive, contiguous, and flat, and to those unfamiliar with the Pine Barrens, it appears uniform. Those who take time to know these places, however, are rewarded with an understanding of the traditions of the Pines. The names of some of the streams are intriguing: Hurricane Brook, Pole Bridge Branch, Stop-the-Jade Run, Jake Branch, Hospitality Branch, Featherbed Brook, Bread and Cheese Run, Ducking Hot Branch, and Papoose Branch. The bogs that are found on these streams were given such names as: Hog Wallow, Ice House, Ragmop, Bert's, Mary's Irongate, Portuguese, Pine Spung, Beaucou, and Wharton. Even the cranberry varieties were given imaginative names by early growers: Howard Bell, Blues Bell, Garwood, Bozarthtown Pointer, Applegate, and Buchalow.

The need for "headland," or upland adjacent to bogs, is the reason for the landscape's extensive, contiguous open space. Cranberry growers need ten acres of upland for every acre of bog, a ratio developed over decades of experience. This guarantees protection from fire, drought, and encroachment by incompatible uses such as residential development. Growers are always on the lookout for new water-supply lands, and monitor local land sales. They rarely sell their bogs, and it is difficult and expensive to develop new ones even on land leased from the state. This landscape has been created and managed by many of the same families over the last century. Mary-Ann Thompson has written about the historic villages and structures (Thompson 1982, 193-97).

> The structures built for the cranberry industry became an integral part of the environment. As the swamps were cut off to set out bogs, the growers used the cedar wood for fences, shingles, siding on buildings, and gates for dams. New impoundments for reservoirs insured water for remaining cedar swamps. The characteristic pine of the area was used for door and window frames, flooring and cranberry boxes and barrels. The buildings in the cranberry villages were spaced in order to achieve the maximum protection from the numerous forest fires. The prime example of integral buildings are the cranberry packing houses (also called sorting houses or screen houses). Originally most cranberry sorting houses were located away from the bogs in town. Cranberry sorting houses are to cranberry villages what cathedrals were to medieval villages. Cranberry villages were centered around the packing house where everyone worked. Along with the screenhouse were workers' houses, tool sheds, storage barns, garages and some-times a company store. Workers' housing were available in multi-family and single cottages. The growers often referred to the housing as camps.
>
> The cottages at cranberry villages also served as housing for the elderly of the Pines. In bygone days social workers feared removing elderly Pineys from the woods to city nursing homes for fear that the old folks would not acclimate well. Retired couples often rotated from plantation to plantation living in small cottages, and, as a result of this custom, they could remain in their beloved woods.

During the first half of the twentieth century, major cranberry communities were basically paternalistic, not unlike nineteenth-century iron-furnace plantations. Bog owners hired a few year-round workers and foremen from the local Piney population and added temporary locals for the harvest. Most Piney workers lived in their own houses near, but not on, the bog-owner's property. The owners cared for the needs of their permanent help and expected loyal service in return. So long as work was done well and basic needs met, owners and laborers had a mutually beneficial relationship.

After World War I, bog owners made increased use of Italian immigrant labor from the Delaware Valley area (photo 11). In the 1930s the famous photographer Arthur Rothstein depicted the wretched working and living conditions of temporary and migrant laborers in New Jersey. A combination of state and federal regulations helped ease these conditions, and more advanced picking methods made most of the unskilled labor obsolete. What remains is a highly integrated, still-paternal system in which little migrant labor is used. The sorting house is still the center of the community, and the owners still live and work at the bogs. Permanent housing is rented or given free to the workers, many of whom are now Puerto Rican, but much of the housing is stucco instead of wood. Some cranberry owners, like the Thompsons, continue to rely chiefly on local Pineys to work the bogs.

How can this traditional land-use pattern continue to work so well? The first answer is that customary private ownership of extensive tracts of land is an effective preservation strategy under certain conditions: economic stability, an exercise of political power by the growers, and a lack of major incompatible development pressure adjacent

PHOTO 11. *Italian Cranberry Pickers, 1934*

to the bogs. Cranberry growing is economically stable because it is technically efficient. Since 80 percent of the marketing is done by Ocean Spray cooperative, growers can be assured of a reasonable profit, which increases their ability to reinvest and remain competitive. Cranberry growers are politically powerful, especially on the local and county levels. In addition, cranberry growers have a major voice in making state agricultural policy.

Unlike other major landholders in the Pines, cranberry growers are not outsiders. Perhaps the secret to their success thus far is the combination of sound business practices, healthy environmental conditions, and an insider's knowledge of the place and its politics. One cannot successfully grow cranberries and be an absentee landlord, because the bogs require too much daily, year-round management.

Development pressure on the fringes of the cranberry district, however, can destroy the landscape. Where no such pressures exist, bog owners see no need for state interference; where groundwater quality and quantity are threatened because of residential development, cranberry growing is jeopardized. Mary-Ann Thompson's bogs, on the western fringe of the Pines, are threatened, and she has reacted by supporting federal and state legislation to preserve the ecological and cultural integrity of the region. To many people, especially those in the environmental movement, Thompson has become the conscience of the community.

Thompson, child of the sixties, former Girl Scout, Wall Street lawyer, and assistant manager and lifelong worker for the Birches Cranberry Company, gave the following testimony to the New Jersey legislature in support of Governor Brendan Byrne's Pinelands building moratorium in 1978:

I am Mary-Ann Thompson and serve as Assistant Manager of the Birches Cranberry Company, started by my great grandfather Martin L. Haines; it has existed for 110 years and we are the largest independent cranberry grower in New Jersey. Our bogs are located in Woodland, Tabernacle, and Southampton Townships in Burlington County.

Before the present moratorium, the development rush put tremendous pressure on our cranberry farm. The Merlino-Yates Bill [the Pinelands Protection Act] gives us a breather by deferring the negative effects of improper planning until a sound constructive plan can be resolved by the Commission.

At Burrs Mill Brook, which feeds 15 percent of the cranberry acreage in New Jersey, there is intense pressure for development. In the preservation area a proposed senior citizen development of 4,500 homes plans to dispose of sewage effluent on a lot adjacent to our cranberry reservoir. Two hundred eight additional homes are being considered by Woodland Township for the swampy headwaters of the stream. In the protection area approximately 38 homes want to drain their runoff into our cranberry canal and into New Jersey's oldest cultivated cranberry bog dating from 1835. Further downstream we find 150 homes being constructed adjacent to one of our bogs. I am concerned that without the Merlino-Yates Bill we will lose the pure water so necessary to cranberry cultivation as a result of incompatible development.

Eighteen months [the length of the moratorium] may be a long time to some but since it will enable us to continue a 110-year-old farming tradition, it is a short time indeed.

Most other cranberry growers considered Thompson a traitor to their conservative political traditions. They said that, by requesting government intervention, she risked destroying the system of private ownership that has preserved the cranberry landscape. She responded that, without government help, the landscape itself would be destroyed.

Mary-Ann Thompson had become involved in a classic battle of point and counterpoint—protection of common resources versus maintenance of individual property rights. The most obvious locus of the battle centered on the arguments between her and her distant cousin, Bill Haines, Jr. Bill manages the largest cranberry bogs in New Jersey at Hog Wallow. His father inherited 400 acres from his father at a time when the bogs produced a crop of 5,000 to 15,000 barrels. Bill, Sr., and his son built up the bogs to more then 800 acres that can produce more than 100,000 barrels in a good year. Bill, Jr., is deeply bitter that people accuse him and other larger growers of selfishness and lack of caring, and he has some right to be bitter—a bad rap, he calls it. Surely, both Bill and Mary-Ann share the same love of their land and even the same vision of the future: to pass on to the next generation a working and harmonious landscape. But they do not agree on the method for doing so. On the one hand, Mary-Ann rightly suspects that local governing bodies would sell for a song the future of her critical communal resources of land and water. It has often happened. On the other, Bill rightly suspects the state of bureaucratic bungling and the taking of property rights that traditionally rest with local people; the state and federal governments have been guilty of this in the past.

Bill is not callous. He and other major growers are the very heart of the cranberry industry. But Mary-Ann remains the conscience because she has a better understanding of the imminence of the threat of change and the need for formal political action. Bill and Mary-Ann need each other: The conscience is powerless without the heart, and the heart is misguided without the conscience. The way in which this point and counterpoint is played out will to a large extent determine the future of the cranberry landscape over the next generation.

THE PINELANDS COALITION

Because of her position as an insider, her legal expertise, and her high level of energy and dedication, Mary-Ann Thompson, became the lightning rod around which gathered the people most responsible for passing the federal and state Pinelands legislation. Around her and her friends was organized the Pinelands Coalition, and it is not clear if the coalition would have achieved its goals without a well-placed insider. While the coalition is a mix of insiders and outsiders, most preservationist organizations were founded outside the Pines and have an overwhelming majority of outside members. Each group in the past has pursued different tactics to "preserve" the region. For example, the New Jersey Conservation Foundation pursued an active and successful program of land acquisition and lobbying, but its outside leaders regarded the Pines as a wilderness ecosystem being continually destroyed by human activities. This is a powerful ideology for recruiting more outsiders, but has little value for bridging gaps among those who use the land for a living. Thompson closed this gap. She was an environmentalist who owned land, a person who could attend cranberry-grower meetings as well as sessions of the U.S. Congress.

In 1978 Thompson and others went to Washington with Representatives Florio (D., N.J.) and Burton (D., Calif.) and officials of national conservation groups including the Sierra Club, the National Wildlife Federation, and the American Rivers Conservation Council, to coordinate amendments to the National Park and Recreation Act of 1978. The Washington work was necessary to get the Barrens on the national agenda and answer criticism based on private-property rights and local control. Congress recognized the Pinelands as a special area, provided funds for planning and acquisition, and directed the state of New Jersey to protect the area. In a public demonstration and press conference at Batsto congressmen, members of the Pinelands Coalition and other preservation groups, and officers of the U.S. Department of the Interior met to announce the beginning of the effort to create a national reserve. Another type of outside control was about to descend on the Pine Barrens.

In 1979, Judi Palumbi of Medford Township, and Mary Tassini, a state Audubon lobbyist from North Jersey, along with Thompson pressed to help State Senator Merlino pass the New Jersey Pinelands Protection Act. They were not crazy environmentalists. Every time an old-guard cranberry grower would lobby against Pinelands legislation, Thompson would speak up for the ''other side of cranberry agriculture.'' On the issue of home rule, Palumbi would talk about the difficulties of running a township environmental-planning program without regional standards or coordination. Tassini knew many people from North Jersey and was an eloquent and persuasive lobbyist. The bill passed, and the state established the Pinelands Commission. Throughout 1979-80 Thompson and Palumbi turned their attention to the Pinelands Commission, but it proved easier to deal with an elective body than a bureaucracy. Both quickly became convinced that the staff, hired by the commission, was uninformed about the Pines. Palumbi and Thompson went to every meeting where consultants reported findings; they constantly called consultants and checked the record of every individual and group chosen to work. They organized comment on the draft plan and in some cases helped supply missing data. Thoughout the year of planning Thompson continually bolstered her friends and supporters in the face of deliberate misinformation, rumors, and fears spread by entrenched real-estate and political interests who felt they would lose prestige, control, and money.

In November 1980 the commission released its plan, which the governor and secretary of the Interior quickly approved. Old-line cranberry interests, developers, some local officials, and some state legislators prepared a lawsuit to challenge the conclusions and cooperated in a legislative effort to amend the Pinelands Protection Act that would have stripped the commission of powers over a wide area. For Thompson, with the battle only partially won, challenge and struggle remain.

Mary-Ann Thompson's and Bill Haines's fight is for a very special place. The Pines generate this feeling, and in those cranberry areas where settlement has occurred the sense of place is strong and families have strong attachments to their communities and landscape patterns. Sam and Caren De Cou live in a lake community that was once an active millpond. The ditches, dams, and spillways of the cranberry landscape have become a residential lakescape and their special place.

Seclusion and the Special Place

LAKE COMMUNITY LIFE

Sam and Caren De Cou live on the edge of Braddocks Mill Pond near the headwaters of the southeast branch of Rancocas Creek. Their home is not far from the divide between the waters of the Wading River, which drain into the Atlantic, and the waters of the Rancocas, which lead to the Delaware River. The house is custom-built on a large lot and set off from neighbors' homes by wide swaths of vegetation. Pines and oaks block almost all other views of other houses and add to the seclusion of the house. There is no lawn. The De Cous understand the workings of the well-drained soils and of the indigenous plant cover that grows under large pines and have made no attempt to import nonnative vegetation. Their one view extends over the lake and down the sweep of the millpond. The house, like those of their neighbors, is set so far back that even looking down the lake in winter, when the oaks are bare, one can hardly see signs of other houses. The only clues of other users are occasional low docks of weathered planks. Their feeling at home is one of seclusion (photo 12).

Despite the sense of seclusion, however, there is an organized community and neighborhood on this lake based on such annual events as the June picnic and the Christmas party, as well as bowling parties and several dinners throughout the year. There are also work parties to drain and clean the lake and to maintain the beaches, the boathouse, and the clubhouse. Thus there are opportunities to ''have your neighbors if you want them.'' The De Cous prize the combination of seclusion and a strong sense of community support.

Quite naturally, outsiders consider the lakes prime fishing and recreation areas, but lake-colony members are emphatic that only members and guests can use the lake and its shores. In the past trespassers, having read a list of names off mailboxes, claimed to have been invited by community members, but a badge system has prevented recurrences, and everyone keeps an eye out for intruders. Patrols are one of the costs of seclusion.

PHOTO 12. *A House on Braddocks Mill Pond*

Caren, born in the city, felt she could adjust to country life, but worried that their young child, Danielle, would have to grow up without kids on the block. Recently some of her fears have been calmed because Danielle has made the nearest neighbors her surrogate grandparents. She makes friends faster than her parents, who still want to stay a little way off from their neighbors. Caren feels that Danielle has made a fine adjustment to life on the lake; she is always outdoors learning through experience about plants, animals, and water. She has playmates at school, and Caren and Sam often drive her to other children's homes, but Caren is surprised to see how little driving she does to provide Danielle with friendships.

On weekdays Sam, Caren, and Danielle drive to Moorestown, which takes about an hour over country roads. Caren works in Sam's insurance office, and Danielle attends Moorestown Friends School, where Caren is a lead parent. Sam is a birthright Quaker whose family has been in crop farming for generations. He went to the same school Danielle goes to and then jumped from farming to financial investments. He serves on various boards and maintains business connections throughout the Rancocas Valley. Moorestown lies quite far down in the basin, and their home lies at the headwaters. Through business connections and family friends, Sam maintains watch over the land economy of the region as well as local and regional politics.

The commute is a long one. With the onset of the energy crisis, Sam and Caren discussed moving closer to Moorestown but decided against it. The family will conserve as much energy as possible in their house and will use fuel-efficient cars, but they do not want to give up their secluded lake home. Sam had lived on farms and in isolated places most of his life. He once tried life in a development, but prefers the woods. When he began life with Caren, he decided to build a house on the spot where he had played and camped as a child; some of his fondest memories are of summers by the millpond. During his childhood the area was almost completely cut off from the agricultural village to the north. The sense of seclusion he feels is also a fondness for times past and a hope his child will benefit from what he had as a boy.

Life at Braddocks Mill is an example of the distinctive life-styles and settlement patterns of the Rancocas River and other parts of the Pine Barrens. Lake communities are perhaps the only kind of development that have really penetrated the forest regions. Although Braddocks Mill and the downstream Centennial and Mimosa lakes represent the upper economic range of this pattern, there is clearly a similar set of aesthetic, recreational, hydrologic, financial, and political relationships in almost every other lake community.

There are, in fact, a surprisingly large number of lake communities in the Pine Barrens, about seventy-five, all created from old millponds or cranberry bogs. Each lake became a focus of a tightly bound community with a strong sense of identity and aesthetics. New residents feel pressure to conform to architectural and landscaping standards and to participate in community life. When local government could not or would not provide services, lake people started their own volunteer fire companies and social clubs.

In Medford, people from lake communities made a major impact on local politics. In the mid-1970s the newly arrived voters elected a rather enlightened township committee with a lively sense of public interest including environmental protection. One of their first attempts was to halt strip commercial development. Through the planning board, residents tried to channel future development into small convenience centers rather than large malls. Old-guard developers and former local officials opposed the plan because they owned most of the property along commercial strips. Township committee and angry lake-community residents, who had not moved to Medford to be confronted by shopping-mall development, forced the planning board to restructure the town plan to meet their desire for decentralized commercial areas.

Lake-community residents remain outsiders, just as do residents of retirement communities. They live in the Pines, but few work there or depend on the area's resources except for recreation. Their friends live in the Pines, but rarely their families. Their vision of the future coincides with a need for seclusion and exclusivity. To lake-community residents, proper land uses include: other lake communities, recreation, berry agriculture, and forestry. The Pinelands Protection Act supported their hope that finally government had recognized and protected their special place.

Chapter 3

EARTH

The soil of the Pine Barrens is not really barren. It is infertile compared to most soils on the west and north, but barren is only a relative term. A family could always wrest subsistence from the sandy ground, but only in certain places and with diligence could one make a living by selling crops for cash. To many people, the earth is as important a resource in the Pines as water or the forests (map 6).

Pinelands farmers are caught in the same bind as many others throughout the country: they would like to continue farming, but economic exigencies may force them to sell their land. Most farmers want to preserve their farmlands, but at the same time they want to retain their right to sell land for the highest possible price, often to a developer. Farmers are important political voices in some sections because they control a great deal of land and the traditional power structure.

Most Pinelands crop farmers established their land-use and settlement patterns in areas where better soils were mixed in with the infertile sands. The so-called shatterbelt (Regensburg 1979) on the northwest is the blending region between the more fertile, clayey, inner-coastal-plain and the less fertile outer-coastal-plain sands. The extensive tertiary remnant, the Bridgeton gravel deposits, south of the Mullica in Atlantic County, and the slight rise of the Cape May formation along the Atlantic coast, inland from the salt marshes, provide slightly reduced permeability and increased fertility.

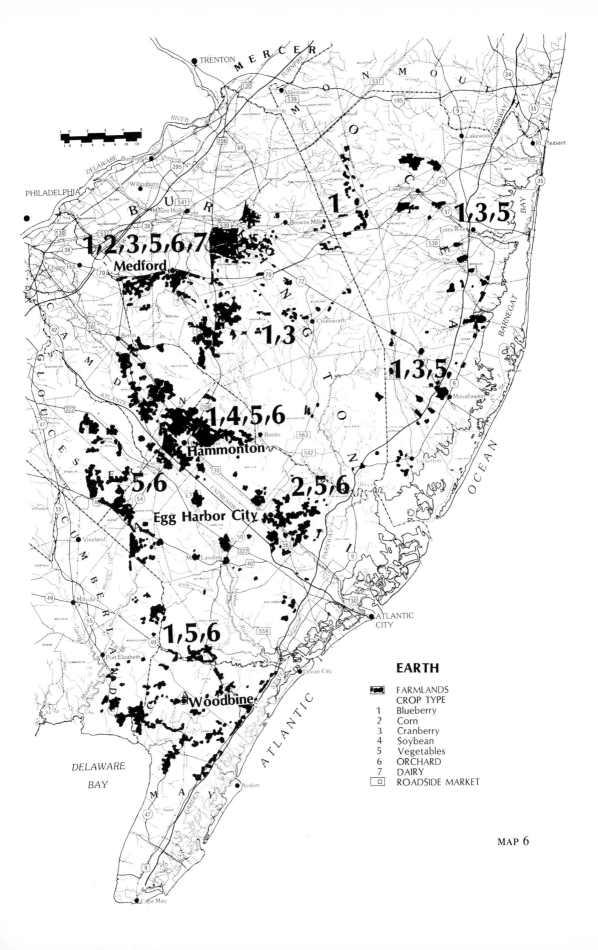

EARTH

	FARMLANDS
	CROP TYPE
1	Blueberry
2	Corn
3	Cranberry
4	Soybean
5	Vegetables
6	ORCHARD
7	DAIRY
☐	ROADSIDE MARKET

MAP 6

Blueberries and cranberries are the commercial crops suited to wet acidic soils, and the Piney garden is the subsistence miracle of organic supplements in sandy forest areas.

The environment, nonetheless, is relatively difficult for agricultural pursuits, and farmers regularly switch crops and develop innovative means of marketing. Unlike baymen, farmers do not work alone. Farm life is more structured and less flexible than the lot of a bayman. Agricultural balance requires constant attention to land productivity and availability, labor problems, weather, changing markets, family continuity, and incompatible adjacent uses such as suburbanization.

Except for some blueberry growers, crop farmers, dairymen, and orchardists do not consider themselves Pineys or members of the woodland culture. For historical reasons they live at the edges of the forest, but their view is outward to the markets.

The farming patterns of the Pines symbolize the general crisis of New Jersey farmers' fight for survival. When one comes to know these farmers, the soil conditions, crop-rotation, labor, and maintenance problems, one can begin to understand their attempts to achieve a balance between traditional life-styles and suburbia—between the forest, farm, and town. Each cluster of fields is an effort to link family, markets, technology, and community. Some efforts are successful, some are not. Some attempts are stressful and others pleasurable. There are examples of both in the Pine Barrens.

Blueberries North and South

TEN ACRES: TOBY GREEN

On a dry day in midsummer, early in the morning, Alfred "Toby" Green is on his way by car, eighteen miles from his home in Pemberton, to his ten-acre blueberry field just west of Chatsworth (photo 13). It will be a good day for picking because the lack of rain in the summer, while it bothers crop farmers, means he can pick the berries without fear of the spoilage that results when the berries are wet. Besides, Green has a good irrigation system and his ground is still moist. He has to work eighteen hours a day in summer to harvest his crop. Dry stretches mean long hours, but good-quality berries and good money.

Toby Green has to pick early in the morning because, once the sun is well up, it is just too hot to work. The Pines and the sand heat up and reflect so much heat that the field becomes an inferno. From noon until evening he cannot pick, but he will start again when it cools off. Green, aided by his nephews, picks with machines developed by small farmers from the area. Blueberry cultivation is specialized and of such relatively little economic value in America that only a small research establishment exists near Chatsworth to develop innovative means of harvesting, planting, and cultivating.

Just before noon Green leaves his field and drives to his mother's home, almost midway between his suburban house and the woodland setting of the blueberry field, where he meets his wife, his young children, his sister, aunt, and mother. His wife has brought the noon meal from home, and everyone eats together. After lunch, in the cool of his mother's shaded garage, the entire family sorts, packs, and cleans the berries. His children, the twins, aged nine, and his daughter, eleven, assemble the cardboard boxes while the rest of the family cleans, packs, and sorts. After the evening picking, the family has dinner together. This routine usually starts the last week in June and

PHOTO 13. *A Blueberry Field in Burlington County*

sometimes continues until the first week in September, although the end of August is more usual. The schedule fits Green's life well because he is a schoolteacher as well as a grower.

Green grew up in Chatsworth, a small woodland village with no more than eighty families. Nowadays most of the residents work for county and state governments and military establishments to the north, although a few of them still work full-time in berrying, woodcutting, gathering, or local commerce. Before World War II most people were engaged in traditional woodland activities, including berrying, sphagnum-moss gathering, plant and pinecone collecting, woodcutting, trapping, and guiding deer hunters, much as Green's father did. In 1947 his father bought the ten-acre blueberry patch, and from that moment on, his five-year-old son, Toby, began working the same ground.

In high school Green became accomplished at baseball, a sport that has a huge following in the Pine Barrens. At Florida State University, he studied physical education, played on the varsity baseball team, and entertained hopes of going into the major leagues, but he returned to Chatsworth to marry a local girl and teach in the local school system. Marriage meant buying a home, and he and his wife moved out of the Pine Barrens to a suburban development near Pemberton, where he now teaches. His wife had worked in the berry fields all her life, and he, too, continued to work berries in partnership with his father. Despite his change of residence and his teaching schedule, Green considers himself a blueberry farmer and a person who has ''grown up in the woods.''

Blueberries provide Green enough income to support his family during the summer and put away some savings for his children's college education. They also provide work the family loves. With enthusiasm Green talks of being at home in the woods and having peace of mind; he is bothered by nobody and nothing when he works his field. He thinks the activity is good for his family. The blueberry patch provides a continual, year-round series of compelling and fulfilling tasks.

In autumn Green cleans up and cultivates. After the leaves fall from the bushes, he begins to prune every weekend throughout the winter. Depending on how much money he made during the season, he may hire help. In spring he hoes and begins his first application of pesticides and fungicides when the weather, temperature, and wind conditions allow it. As soon as the buds blossom, he rents beehives, and then waits until picking time in June. Green knows every bush and variety in his field. Good berries require great care, and he says that "ten acres is just about the size one person can manage."

The Green family has no time left to market their own product, so they depend on the TRU BLU cooperative in New Lisbon. Like the machines developed for harvesting and the varieties developed for market, the small local growers helped develop their own marketing organization. Under the 1928 Kapper Volstead Act, farmers can form marketing cooperatives free from antitrust laws, and cooperatives are common throughout the mid-Atlantic region. Unlike farmers in other coops, however, more than half of TRU BLU's members are small holders, all of whom live north of the Mullica River. This is the region of least disturbed Piney culture characterized by traditional life-styles.

TRU BLU has 88 members, 48 of whom have holdings of about ten acres, and 40 of whom are full-time, large-scale growers. The coop produces 20 percent of the state's total production with small growers producing about 25 percent of the coop's amount. The coop is, itself, tied closely to the development of the American cultivated blueberry industry that began in the Pine Barrens with the propagation of varieties derived from two indigenous species, *Vaccinium australe* and *corymbosum*. Frank Colville, a U.S. Department of Agriculture agent, made the first crosses of wild plants in the northern Pines with the assistance of Elizabeth White of Whitesbog, who sent knowledgeable gatherers to wild-berry patches in the first decade of this century with gauges to measure and baskets to collect only the largest and hardiest specimens. To these White added other varieties from New Hampshire. Local people, in cooperation with the staff in Chatsworth at the Cranberry and Blueberry Research Station of Rutgers University, propagated the original three varieties, and from the Pioneer, Cabot, and Katherine came thirty-two others (Applegate 1979). In 1916 White planted the first cultivated berries at Whitesbog, and in 1925 Rutgers established the cranberry and blueberry laboratory which still exists.

Well-established cranberry growers started the TRU BLU cooperative in 1927. These families included Budd, Haines, White, Reeves, Lee, and DeMarco, but some, like the Thompsons, never joined the coop, claiming it was monopolistic; coop members claim that those not in the coop get the same benefits without paying their dues. With capital from cranberry operations and large holdings in the major water-sheds, larger growers developed substantial commercial operations. The process was predictable, since cranberry bogs and blueberry fields are handmaidens; the former are on water-logged soils, the latter on damp soils directly adjacent to the bogs. Some workers, like Toby Green's father, who learned their skills on larger fields, bought their own land and joined the coop as well.

By 1932 blueberry production used only 170 hectares of land in the Pine Barrens, and in 1963 the amount had risen to 3,400 hectares even though the number of farms declined from 501 to 383 in that same period. A cost-price squeeze made many small owners abandon their plots, as prices for berries remained constant, while labor costs increased quickly (Applegate 1979). Mechanization, however, helped the industry recover, and by 1967 growers had invented their own mechanized harvesters. High labor costs have much less effect on people like the Greens because they use their whole family as workers.

The 1970s saw a slight increase in the area of production but not in the number of farms. The old founding families continued to buy up small plots, while small growers maintained their equilibrium. From 1969 to 1979 the blueberry industry's gross annual value was $8 to $14 million, with TRU BLU accounting for $1.5 to $3 million of the total.

Harry "Junie" Bush is the TRU BLU manager. Descendant of a lifelong farming family, he has spent many years marketing berries. From the modest coop building, where he has only a desk and phone, he compiles weekly information on worldwide berry markets. Weekly he receives reports from the American Institute of Food Distribution, the *Washington Food Report*, which covers legal matters, the *Weekly Digest*, a summary of magazine and journal articles on growing and marketing trends, and *Food Markets*, a publication that follows week-to-week prices, crop reports, and USDA crop projections. In addition USDA sends Bush the *Federal Cold Storage Report*, a monthly summary of where and how food is being stored around the nation. With this information, Bush estimates the time of market arrival and the quality and quantity of berries arriving from other domestic and international berry regions. He augments this information with phone calls to commission merchants and food brokers all over America.

Commission merchants, who have exclusive rights to TRU BLU berries, represent the coop in New York, Philadelphia, Boston, and Baltimore, while Bush deals with a variety of food brokers in the West and Midwest. He markets the bulk of the berries in the Northeast, Midwest, and Canada. When in 1979–80 Bush noticed a potential glut of berries for processing as inventories of frozen and canned berries rose, he informed the coop's board of directors of his findings. The board told the membership, who responded by switching to different picking methods for the fresh market. Because individual consumers look for big, plump, unblemished, highly colored berries, picking and packing must be done with an eye toward quality rather than high productivity. Prices for fresh berries have been excellent, but the demand for high quality required picking expertise and close supervision of sorters and packers. Small- and medium-sized growers produce the highest-quality berries because they can enforce better quality control. Mrs. Green, working with her family in her mother-in-law's garage, has done a job equal to or better than fifty workers bussed from Philadelphia to a modern packinghouse. The majority of small, highly skilled, and dedicated growers in TRU BLU's membership allowed the coop to increase by 92 percent the amount of berries sent to the fresh market in 1979 and 1980. This extraordinary level of cooperation resulted in excellent returns for the growers, so for the winter of 1980 Toby Green could hire a pruner.

The TRU BLU cooperative is an outstanding example of the way Pinelands residents achieve a balanced, productive landscape at the edge of suburbia. There are, however, other ways to work the blueberry landscape. South of the Mullica is the area of Italian settlement where blueberry farms can be enormous. These large farmers have balanced the same forces for change and stability in different ways from those developed by their colleagues to the north.

TWO THOUSANDS ACRES: DUKE GALLETTA

Early on a July morning, Duke Galletta and his son Al sit in the office of their packinghouse and market their fresh blueberries all over the United States. July is the time of the "glut," and the Gallettas have to "move" over ten thousand flats of berries every day during the peak season, or they will not make a profit. Their office has two metal desks, two phones, adding machines, a calendar, a map of the United States, and books listing food brokers throughout North America. With over forty years' experience, Duke knows how to "move" berries.

As the harvest proceeds, Duke sets a weekly price that he decides on Fridays. Over the years he has built up contracts with food brokers in many places representing chains of food stores. The brokers call Duke every morning to place orders, but if he cannot keep pace with the orders, Duke gets on the phone to the brokers. He prefers to ship to the fresh market, where he gets the best price, but will send consignments to the "bulk house" for freezing and canning.

Duke has an extraordinary feel for the market from past experience and knowledge of blueberry production. He even tries to match the readiness of his berries in advance with the desire of a chain store to have bargain days, and he lines up "sales" throughout the season. Over the years he has learned where to ship berries and has concluded that people "won't eat cultivated berries unless they have had experience with wild ones. If an area has wild berries, people will buy fresh cultivated berries." For this reason, although he has tried to open major markets in California, he has so far not been very successful there.

Duke's map of markets shows not only a partial distribution of the wild blueberry but also the diffusion of berry foodways across the United States and Canada. He is always thinking about ways to expand his markets and believes that, if he can increase production and markets, he can maintain his profit level despite rising costs of labor, fuel, pesticides, herbicides, and fertilizers. He is now looking to foreign markets like Japan, China, and Brazil, where blueberries are considered an exotic delicacy, much like kiwi fruit is here. When he was president of the North American Blueberry Council, he helped guide it to explore foreign markets and new products like wine, sparkling wine or Blueberry Duck, and fruit drinks from processed berries. The popularity of yogurt, for example, has had a major impact on demand for such berries, just as, in the cranberry world, the acceptance of cranberry juice saved part of that industry. Competition does not really worry him because he has a secure market position; he has been in blueberries from the beginning of commercialization and has been a leader in the development of the many techniques of cultivation, propagation, breeding, and marketing. South Jersey is a benign environment that steadily, with only rare interruptions of storms or drought, produces high-quality berries, while competitors in more northern areas often suffer frost and hail damage.

Duke is now ready to retire, and he and his brother will pass the business to their sons. Next to his modern, split-level home are the houses of his married daughters and a vacant lot for Al, who just married. On his desk are brochures for Florida and Caribbean vacations. He has enjoyed his life's work, but things were not always this way.

In 1914 Duke's father moved the family from South Philadelphia to Hammonton because he wanted his children to grow up in the country. The father worked as a foreman in a Philadelphia woolen mill to which he continued to commute by railroad for over thirty years while Duke and his brothers went to Hammonton High School and began to work summers on the surrounding farms. Duke and one brother worked for the Collins family of Moorestown at the Atlantic Cranberry Company southeast of Hammonton. In 1932 the brothers hand-grubbed the first sixteen-acre Galletta blueberry field. To stay in business they did all manner of seasonal work from cranberry harvesting, to pruning, to woolen-mill work in Hammonton. Always they saved money to buy small machines for their little farm, gradually adding more cleared ground each year until, by 1950, they had over one hundred acres.

In the postwar period everyone was thinking big. Duke, like many others, wanted to expand, and the small fruit industry offered promise. The Atlantic Cranberry Company went up for sale because the Collinses had no children to continue the operation, and Duke was the obvious person to sell to. He knew the farm and berry business and wanted to stay in the berry business, but not in cranberries. He reasoned that, although there had been extensive cranberry cultivation in Atlantic County, the possibilities for expansion were small for both economic and environmental reasons.

Duke believed that the swamps south of the Mullica were ''deep swamps, harder to convert to productive bogs than those around Chatsworth where there's savannas, land that gets gradually drier.'' He felt there was not enough water for expanded cranberry operations: ''The streams are okay, but you can't buy land to protect the watershed. Up north they have oceans of land.'' Furthermore, blueberries require less starting capital, and, when Duke got into the business, blueberries looked like a better investment because of market conditions.

Soon after purchasing the 1,700-acre Collins property, the Gallettas changed its name to the Atlantic Blueberry Company; they ripped out miles of intricate cranberry systems and cleared the forests protecting the bogs. They plowed under 60 acres of old blueberry varieties and planted newer ones. In all, over 400 acres of cranberry bogs were filled, and since 1950 the Gallettas have increased the size of the blueberry fields every year.

Duke's marketing techniques have changed as drastically as his production and his company's size. From 1937–46 he shipped to a New York commission house, and from 1946–55 he sold to the Hammonton Fruit Auction, where his berries went on the block to the highest bidder. Sometimes the auction had too many berries and too few buyers, so Duke decided to go to direct sales. Since 1955 he has dealt with brokers who provide their own transportation.

In 1980 the Atlantic Blueberry Company was run by Duke, his three brothers, and the wife of a deceased brother. The sons of the second generation actually do most of the work. The company has about 70 year-round employees, most of them pruning in the winter and picking in the summer. During the picking season, the Gallettas employ 150 migrant Puerto Ricans who live on the farm; the same workers return annually and are only replaced by word of mouth. They also hire between 1,500 and 1,800 day-haul laborers from the Delaware Valley and the Vineland and Atlantic City areas, many of whom also return every year. One hundred local high-school students work the processing house. At the height of the season, this work force picks and packs nearly 12,000 flats daily. Besides berries, the Gallettas also sell bushes, between half a million and a million each year. They are dug and shipped during the dormant season and can be shipped "bare rooted" to Europe and Asia.

Duke is a well-respected, well-connected member of the region's agricultural community. He belongs to the New Jersey Farm Bureau, the United Fresh Fruit and Vegetable Association, the North American Blueberry Growers Council, the Hammonton Rotary, and the Chamber of Commerce.

Hammonton is the name of a village, a municipality, and a general agricultural region that was settled first by New Englanders and then by southern Italian peasants. It was transformed from a forest reserve for rural industries to a prosperous agricultural area. Since their arrival, Hammonton Italians have made their money from fruits and vegetables first and small industry second. All family members work in the fields. There is a Hammonton story of two Italian children who heard their Anglo classmates were going on vacation. The younger child asked the older one, "What's vacation?" to which the older one replied, "I really don't know, but I think it's someplace in New York." These children knew that as soon as school ended, they would be in the fields.

Today Italian families in their fourth generation on the same land have expanded their holdings and own most of the large farms in the region. Russ Clark, the former mayor of Hammonton, owns some of the most productive uplands in the Pinelands.

The Italian Orchards of Hammonton

FRUITS AND VEGETABLES

Russ Clark's home and his Green Mount farm stand sit on the crest of a small rise along Route 54, just west of Hammonton's town center (photo 14). From that crest Russ and his family can look out over three of his orchards to his packinghouse and down the road toward his relatives' houses and his original homestead, a three-story frame building now occupied by a law firm. From his yard, Russ, like Albert Reeves, can look out over his life as well as his lands.

Clark owns 226 acres of prime land and leases another 100 acres. When he was a freshman in high school, his father gave him his first plot of land and in 1956 gave him his own farm. Following this tradition of stewardship and paternal help, Russ hopes to finance loans for his sons, who will then assume the mortgages for new blocks of land, thus increasing the family holdings.

Russ considers himself lucky that he was not only born into land, but married into it; both he and his wife have relatives in the Macri family. On his marriage, Russ inherited land passed down through his wife's family after which he bought land from grandfather Macri's estate. From an uncle Clark leases a farm for $200 a month, and, on the uncle's death, he will become owner of that land. Although the Macri family holdings form the core of Clark's best land, he is always on the lookout for available land to increase production and, thus, keep ahead of inflation. Clark is aware of the whole farming situation in Hammonton—the status of productive and nonproductive land, the viability of family farms, and when the next parcel of land might come on the market. Purchase of land from an estate sometimes requires as many as a dozen or more

PHOTO 14. *Aerial View of Hammonton*

signoffs from scattered relatives before title can be cleared, but land that cannot be bought can possibly be leased.

When Clark looks out over the fields, he sees a land in flux: the change of seasons, shifting ownership patterns, the consolidation of his own holdings, the constant pruning and renovation of the orchards. All this is common knowledge in Hammonton where "everyone knows everybody else and everyone knows everyone else's land." But title to land is only part of the formula for prosperity; Clark believes that "once you got it, you have to use it in the best way possible." Stewardship and constant acquisition precede the necessity to nurture fertile ground.

Russ Clark's farm is larger than the average one in his farming area of western Atlantic and eastern Camden counties. The county agricultural agent, Charlie Dupras, with whom he is friendly, estimates that in the immediate vicinity of Hammonton, Folsom, Mullica, and Hamilton townships, there are about three hundred farms with an average size of seventy-five acres. In an expanding economy, good agricultural land for fruits and vegetables is scarce, and one cannot always look to buy new land to increase productivity. Thus Clark and his neighbors manage intensively what lands they have. On the sandy coastal plain, underlaid with plentiful groundwater, he irrigates every acre and seeks to enrich the soil with ground-up brush from pruning, old leaves from the city's parks, and municipal sewage sludge. Small rises and hills receive the most intensive organic supplement because they have the excellent cold-air drainage necessary to protect crops and trees from spring frosts.

Not all land in Hammonton is "good," meaning fertile. If Clark could farm every acre he owned, he would have more than two hundred acres in peaches and thirty-five in apples, but he cannot. Over time the Hammonton fruit growers have found that "sugar sand" will not produce anything because it is barren, white ground on which only scrub oak and pine grow. It is found ribbonlike along ridges, a remnant of yet unknown origin, but possibly of ancient sand dunes. Because it is discontinuous and spotty, the soil scientists at the Soil Conservation Service do not map these deposits, but local farmers know where they lie, and newcomers, who thought they had found good, cheap farmland, have learned the hard way.

Upland soils in agricultural sections of the Pines are generally difficult to know intimately because minor changes occur in them from spot to spot. In most areas, certainly in the Midwest, the SCS has an easier time mapping soils because they are more uniform. What one sees on the soils map is what one gets in the ground. This is not so in the Pines, where, because soil fertility is low or moderate in any case, minor changes make necessary major agricultural shifts from apples on sandy loam to peaches on loamy sand.

Soil is a living medium. A good farmer husbands soil just as a rancher tends livestock. The base material may be inert, but that material is so mixed with plants and bacteria and dead organisms and fungi and worms that the medium, under a microscope or in the hands of a farmer, does indeed come alive. Soils must be fertile to grow crops, and fertility means getting the proper amounts of nutrients, like nitrogen or phosphorus or potassium, to plants at the right time of year. Lettuce, for instance, needs much more nitrogen than phosphorus or potassium, while the opposite is true of sweet potatoes and other root crops. One can easily tell the fertility of a soil by its color and texture: dark and crumbly if it is rich, but gray and scratchy when not. A soil with too much clay, although the clay is rich in nutrients, becomes so compact and sticky that it will not allow much air inside, and it becomes waterlogged and suffocates or drowns most plants. A sandy soil, on the other hand, has plenty of air and is well drained, but it will not hold nutrients or water. The right kind of soil, therefore, is a mixture of large-particled sand and small-particled clay or silt. This mixture is called loam, or, colloquially in the Pine Barrens, "loomy ground." Much of Russ Clark's success lies in his intimate knowledge of changes in soil properties, not only on his own land but almost everywhere in the Hammonton area.

The larger, more established families—the Macris, Curcios, De Augustinos, Wuillermins, and others—bought or cleared large tracts of the best land. Later, less affluent arrivals had to settle for smaller and marginal sections which, in earlier times when labor was cheap, could produce blackberries or "black diamonds." As the price of labor rose, the cost-price squeeze that affected all small fruit producers forced the abandonment of black-diamond plots. Now along the the backroads there is a patchy pattern of low, white rises in twenty- to thirty-acre blocks, swept by wind and covered with Indian grass and some red cedar, oak, and pine. Small, well-kept homes border the fields, and in back may be a thin woodlot. According to Clark, ''No one is going to grow anything on this dead ground. If they could, someone would have bought it and planted it a long time ago.'' He once commented that the gas company offered him acres of sugar sand, but he said, ''I wouldn't even mow it for a fee.'' To Clark's practiced eye, these variations in land are quite distinct, but the less well-informed observer must rely on soil color, for the white lies distinctly on top of or adjacent to the more productive yellow-brown soil.

When Clark looks at his orchards, he says, ''The only thing that stays the same is change.'' Nothing on his landscape is static. His family has been in trees since 1900, and he knows that no orchard is permanent. A peach orchard has a life of twelve to fifteen years, while apple and pear orchards can last up to sixty. The lifetime of any one area varies by tree variety and soil type, and Russ constantly mixes crops and changes the complexion of his fields.

Ninety-eight percent of Russ's trees produce fruit that suits the taste, or more likely appeals to the eye, of the general public. He has two or three scattered acres of experimental trees that he grows to his own taste or to test a new variety's adaptation to local conditions, but basically he grows haven-type yellow peaches and five varieties of apples—Macintosh for summer, Red and Yellow Delicious for late summer, and Stayman Winesaps and Romes for fall, plus some Jersey Reds for processing. Buyers get used to the looks of certain apples and will not buy any other at a certain time of year. Clark tried some Mollies Delicious apples for a while; they are Delicious in shape and red in color, but they ripen in summer along with Macintosh, and most people will not buy Delicious-type apples in summer. Another new variety, Spigold, is a wonderful, sweet-acid, large, fine-keeping apple, but it is neither red nor yellow, and consumers will not buy it. Similarly, many people will not buy a peach with white flesh, no matter how ripe and luscious it may be, so Russ used to use white peach varieties for pollenating his yellow peaches and for home eating.

Clark creates and experiments with many combinations of crops and trees. He may take out old orchards and plant totally new ones or just replace some varieties. He will mix crops—vegetables with fruits. If cucumbers do not do well in spring, due to cold rain, he will switch to cabbage, a 120-day crop. He may double-crop a field: If he can get his cucumbers out early he can put in a late field of beans, pickles, or late tomatoes that he can harvest right up to frost in November. If he has a large fall apple crop, he will not plant late vegetables, but if his peaches are light, he can ''go heavy on the cucumbers.'' Any mixture is a balance between weather, soil productivity, available labor, and crop conditions. In every case he shapes the produce like clay in his hands.

For every change in the natural environment, the economic climate, and the political arena, Russ Clark must find a way to maintain the viability of his operation. Since 1956, he has dealt with rising labor costs, labor scarcity, and the threat and promise of suburban development. He has often said, ''If farming remained feasibly profitable and the labor was ample, a farmer would not sell out.'' But when he thinks about labor problems, he shakes his head: ''No one wants to see a hoe, and now the Atlantic City casinos take all the labor.'' To deal with the problem, he has radically changed operations, using new chemicals, tools, and procedures to reduce as much as possible his reliance on hand labor and, to some extent, fuel.

Clark has recently gone to bulk bins, bulk dumpers, forklifts, and bulk trailers to eliminate hand labor, and, for the same reason, he uses hydrocoolers instead of ice for cold storage. He dilutes sprays and uses short-residence sprays that take less fuel and fewer man hours. Clark irrigates every acre with self-moving systems and is interested in wind power to move water with less human effort and fuel. He plants his trees closer together and uses dwarf rootstocks to eliminate ladder work and provide workers with shorter rows. Finally, he grinds his brush and prunings into the soil, adding nutrients and saving money.

By introducing these procedures, Clark has cut back significantly on labor requirements and now hires only three year-round workers, although he still needs a large number of migrant laborers and local schoolchildren for picking and packing. He is unsure if he can count on his children for help. Unless his daughter marries a farmer, she will not be involved in the operation, and one son is too young to work in the fields. The other son attends Delaware Valley Agricultural College and prefers the marketing end of the business. This son's interest in sales provided the impetus for building the Green Mount Farm Market next to Clark's house. The market provides welcome income, but most importantly, together with the farmlands, it provides a means of assuring family and business opportunity for young people, thus assuring family and social continuity.

Clark's mother used to have a small vegetable stand next to the house, but Clark's son convinced him to build an enclosed structure for year-round sales. By its third year of operation, the new stand made a good profit on a full range of nursery products and Christmas trees as well as fruits and vegetables. In spring flowers and shrubs go on sale, after which the family's strawberries come to the stand in June, followed by the whole range of fruits and vegetables Clark grows. He can get double the price retail that he can on the bulk price of fruit and vegetables and still keep the price competitive with supermarkets. Year by year the volume has gotten greater; now the Green Mount Farm Market draws from the shore trade (people going to and from the beaches) as well as from locals who know they can find top-quality produce.

The little produce market with its clientele is a miniature of the Hammonton-Vineland-South Jersey agricultural region. Bordered on the east by recreational shore communities, on the west by suburban fringes, inhabited by numerous produce brokers from national supermarket chains, and served by the Vineland food auction, the region enjoys easy access to national and regional markets. Clark has never had a marketing problem. "Whatever I can produce," he says, "I can get rid of. We are right in the middle of the marketing areas." He produces for the tastes of the mid-Atlantic region and has benefited from the growth and development of South Jersey, even though that same growth has created labor shortages.

Russ Clark derives direct benefits from Atlantic City's growth and the increase in traffic between Philadelphia and the casinos, but he has not, up to now, suffered from the competition of suburbs for agricultural land. This is not mere happenstance, because Clark was mayor of Hammonton and served on the town planning board. "It galls me," he says, "to see prime land go. I would cry to see a bulldozer come in and put in homes. I show my sons the fields so they know their value." Control of suburbanization is essential for a healthy agricultural society. The irony is that Hammonton farmers depend on suburban markets; on the other hand, they see the suburbanization of Hammonton as a threat to their livelihood. The siting of one residential cluster near an orchard means all agricultural activities must be scheduled so as not to interfere with residents, and the construction of homes might well occur on good land. Thus Clark, as mayor and planning-board member, had helped promulgate strict regulations to control development and to prevent what he calls another Levittown. The regulations, coupled with judicious placement of water and sewer lines, have prevented ill-advised development.

Clark is not against development. He looks at it in economic terms and has decided that people can make a lot of money from orchards, yet he believes that Hammonton needs new tax ratables. In his view, the dead land of ''sugar sands'' can accommodate enough development for many years to come. Some of this development can provide new homes for casino people, and some can be used to house married children who want to live near their families.

Family and community are important in keeping agricultural life viable in Hammonton. Families are the underlying force behind the assemblage and care of the land to provide future homes for children and grandchildren. The Clarks, Macris, and others save marginal lands for their daughters' homesites, while the good land is generally passed to the sons. This process has created elongated clusters of houses along all the roads that cut through the large orchards and fields of Hammonton. Homes for laborers are concentrated at packinghouses or other work sites, but family members live within visual contact of each other up and down the road network.

Just as farmers donate land to their children, they lend friends and neighbors equipment. About once a month Clark lends his planter to someone or lets a farmer use some of his cold-storage space. Renting land is a form of exchange that helps people pay taxes and keeps land mowed and open. He will also lend people his workers if a pressing need arises.

As mayor, Clark received calls from everyone. He was elected twice and benefited from that position, but still pays the price of lack of privacy. A CB radio in his home keeps him in constant touch with a full range of problems—from fires to gypsy-moth attacks. People call his home, and his wife relays messages to him over the radio. He and his wife give constant care to family, friends, and neighbors.

An Ethnic Archipelago

We are about to meet three families who work the land, one for cash, the other two for family use. They live within a ten-mile radius of one another, but none of them knows of the others' existence. Although they come from different backgrounds, unseen and unspoken bonds tie them together: a love and understanding of the soil, a strong sense of independence, commitment to to their families, belief in the sanctity of private property, and, in two cases, loyalty to church and community. In this they are not different from the Gallettas and the Clarks, but none of the three holds the political and economic power of the Hammonton farmers. Rather, they represent hundreds of other families whom one can find in the southern Pines.

Historically, farming and gardening have been a way of life and the chief source of income or food for an untold number of people in this southern section. They are a critical part of the social and ecological equilibrium in the Pine Barrens and contribute to social and economic self-sufficiency, regional economic balance through use of agricultural markets, sensitivity to ecological potentials and limitations of the land, and habitat diversity for wildlife. Farms and gardens are also an integral part of the visual landscape of the Pines and the source of some of its finest examples of vernacular architecture.

As Elizabeth Marsh (1979) has indicated, three elements contributed to the contemporary landscapes of Hammonton and other southern areas. First, because forest industries in the Pines were dead or dying by 1860, land was comparatively cheap, and extensive holdings were for sale. Second, in the 1850s and 1860s a railroad net developed, not only between Philadelphia and Atlantic City, which had two competing lines, but down through the Great Egg Harbor and Maurice River water-sheds; only one line was built north and south through the Pines and east-west in the northern section. Last, commercial fertilizers and irrigation systems came into general use after the Civil War. An added element was the great European immigrations of the late-nineteenth century to the East Coast. Thus four concurrent events helped create the landscapes of the southern Pine Barrens: cheap land, a modern transportation system, new fertilizers that redefined the fertility of otherwise meager soils, and the cheap labor of immigrants.

In effect, settlement patterns of the new immigrants resemble Marsh's ethnic archipelago:

> The first new European settlers [after the Yankees], the Germans of Egg Harbor City and New Germany (now Folsom), arriving in the 1850s came from middle-class antecedents. The money that built Hammonton came with its pioneer gentry from New England in the 1860s. The famous Jewish colonies, those established at Alliance and in the Vineland area in the 1880s, and Woodbine founded in 1891, were supported during their first years by the Hebrew Emigrant Aid Society, the Baron de Hirsch Fund, and similar resources.
>
> Once Communities were established, sequential occupation began. Out of their prosperity the Yankees and Germans recruited Italians to do agricultural and railroad work. Northern Italians came to Galloway, near Egg Harbor, to work the vineyards. Hammonton has long since become an Italian town settled chiefly by people from Gesso, Sicily and the province of Naples. This migration began in earnest in the 1870s. And now the sequence continues as Puerto Ricans fill the labor niche once occupied by Italians.

For a hundred years, from 1850 to 1950, the ethnic archipelago was a cultural refuge for Germans, Italians, blacks, Puerto Ricans, Eastern European Jews, Russians, Ukrainians, and Anglos. It is one of the few places left in the East where sons and daughters of the pioneers are still alive, people whose parents and grandparents first grubbed stumps to clear lands.

A few refuges in the Pines, Hammonton in particular, were so successful that they continue today as coherent, vital ethnic communities. Their success lay not only in internal factors such as religion, language, and family, but in luck, background, choice of land, and their people's ability to create a balanced economy by exploiting many resources, switching from one to another much as Russ Clark has done.

Most refugee areas died or lost their identities, such as those of the Russians and Ukrainians, some of whom were assimilated and most of whose children drifted toward the cities. The Jewish settlements are now only a memory in Woodbine, whose magnificent brick synagogue is a national historic site, bereft of the ten Jewish men required to conduct services.

GERMAN TRUCK FARMS

Egg Harbor City is in transition as the cultural hegemony of its German majority has dissolved and its agricultural fields have reverted to secondary growth. Sitting in his ranch house, two hundred yards back from Mannheim Avenue, cornered by woods on the northwest, Willard Grunow mulled his future in farming. His son worked for the Farmers Home Administration and was making too good and secure a living to return to the vicissitudes of farming; his daughter, a senior at Wellesley College, was off in the Mojave Desert on a geological expedition sponsored by MIT. She was not a likely candidate for farm life. Willard complained about land values and labor problems, as do all farmers, but it was really the future that he viewed with hesitancy. The farm, its machinery and land, furrows, sprouts, and memories would ultimately go to someone outside the family. Full value for the land would ease some of the anxiety—$10,000 an acre instead of $5,000—but still, selling the property was not like selling stocks or bonds. Had Willard bought stock after he returned from the Pacific in 1946, he might have become rich on all kinds of investment opportunities. The truth be known, Willard, like Toby Green, wanted to be a professional baseball player more than anything else. He had the size, and the throwing arm of a catcher, but he could not hit. The Boston Braves sent him out to Oklahoma to try out for one of their Triple A farm clubs, but he could not make it. That dream shattered, he returned to the family farm in South Jersey and bought land just east of Egg Harbor City that the town of Galloway owned.

Galloway Township, like most towns in the United States in the 1940s, owned a lot of land that had reverted to the tax rolls during the Depression. In Galloway that land was going for $200 for a twenty-acre parcel. Willard bought two parcels and, like his father and grandfather, settled into farming. His father had been killed by a drunk driver in 1931, when Willard was five. It was a bad year for his father to die, for any father to die. At the time he raised classic South Jersey crops—tomatoes, beans, white and sweet potatoes, peppers, and eggplant, depending on the market and his forecasts. During the next few years Willard's mother raised lima beans for some cash income, since limas are a sure crop and easy to pick. Meanwhile, aunts and uncles pitched in to help the family, which scraped through the Depression like many others in the region.

The family farm on which Willard grew up was started by his grandfather, who was born in Egg Harbor City. Grandfather Grunow moved several miles east from Egg Harbor at about the turn of the century to create on Mannheim Avenue a farm out of the old woods that had once been fuel for the iron furnaces. As the family grew, the sons and their families bought more land along Mannheim, until the street became essentially the Grunows' street, just as Cologne Avenue is the Liepes' and the Roesches' street or Moss Mill is the Sahls'. In fact, almost all the agricultural area east and west of Egg Harbor was settled in twenty-acre blocks by the German immigrants and their children who began Egg Harbor City in the mid-1850s.

At the age of thirty-eight great-grandfather Grunow and his thirty-one-year-old wife arrived in Egg Harbor sometime during that decade. Willard did not know where they came from originally, and it would be hard to discover whether they came straight from Germany or had first settled in another eastern city, most likely Baltimore or Philadelphia. Egg Harbor was established as one of the first islands in the South Jersey archipelago. The railroad and several Philadelphia Germans created it in 1854. The Philadelphians—the brothers Schmoele and Mr. Wolseiffer—bought almost 40,000 acres between Hammonton and Atlantic City and organized the Gloucester Farm and Town Association. (Gloucester referred to the old iron furnace, which had operated on the northern edge of the town, near the Mullica River, in the first quarter of the nineteenth century.)

What the original town fathers wanted was written in an 1859 advertising pamphlet, aptly entitled *"Was wir wollen"* ("This Is What We Want") (Cunz 1955?, 11).

A new German home in America. A refuge for all German countrymen who want to combine and enjoy American freedom with German *Gemuetlicheit*, sociability and undisturbed happiness. A place to develop German folk life, German arts and sciences, especially music. A place around which we can build German industry and commerce, a practicable harbor and railroad connection to all parts of the country.

Harassed by the anti-immigrant sentiment of the Know-Nothings of the period, Germans settled Egg Harbor and made the town work. It was incorporated in 1858, and people began immediately to buy stock in the association. Between $300 and $400 bought a twenty-acre farm plus a town lot 100' × 150' in the town center, while $78 bought a town lot only.

The association began a successful advertising campaign to lure Germans and German-Americans to Egg Harbor, often relying on anti-immigrant sentiment to make the point. Witness an 1858 advertisement in the Baltimore *Weeker* (Cunz 1955?, 12):

You had courage enough to leave your beloved fatherland, to escape the hands of the tyrants and to search for a new home here. In Baltimore you found all this for only a short time, since during the last years neither your life nor your property has been safe. Are you chained to the soil? Did your former courage evaporate? Do you want to be robbed and killed? No. Leave a city in which you and your children are threatened by misfortune and contempt. Thus I invite all of you who still have some courage to leave this town of robbers and murderers. Come and be informed about a new free home.

The association's founders laid out a classic American grid-patterned town with German quirks. All residents had rectangular lots, but every block was bisected by an alley. Neat gardens stood in the back of every dwelling, and each garden faced the alley, across which neighbors gossiped or junk accumulated, so at least the facades of all streets were clean—no gossip or junk allowed in public. Further, Egg Harbor's townscape is highly integrated, in terms of both economic activities and social class. The brick and stone banks and town hall bear the burgher stamp of moderate wealth, and the rambling Victorian wood townhouses have character without ostentation. These moderate mansions are scattered throughout the central part of town next to smaller houses, some even across the street from small clothing or cleaning industries or kitty-corner to local drinking establishments like the Liberty Beer Garden. What appears to the stranger as just another American small town is really an image of a German *Freiestadt,* one of the free cities of Germany that had their own municipal governments run democratically by burghers.

For seventy-five years the Egg Harbor Germans sustained themselves, despite World War I and the problems it created for German sympathizers; although Egg Harbor was tied to the regional economy, it was sufficiently isolated geographically to escape the brunt of anti-German sentiment. The town thrived on a mixed and balanced economy of commerce, industry, and agriculture. It was, until Prohibition, the viticultural center of South Jersey and the Delaware Valley. Prohibition and the Depression killed the wine industry, and postindustrial changes after World War II made most other industries obsolete. Young people no longer felt proud of German culture, and the German language, once Egg Harbor's official tongue, disappeared from its schools. After World War II, then, Egg Harbor lost much of the flexibility of its economic base and its intense cultural pride.

Willard Grunow was caught in this long process of decline. He married Loris Kienzle, the daughter of a neighboring German farm family, but much of her life is now spent outside proscribed limits of German dicta—church, children, and cooking— because she is a curriculum coordinator for the Galloway Township schools and a woman of some sophistication. Will and Loris are rightfully proud of their worldly knowledge, but the cost has been further erosion of the old German community.

Willard will continue potato farming until he retires. In the fifties and sixties he worked a variety of crops, like eggplant and peppers, and even formed a cooperative to market sweet potatoes with other German farmers and a few outsiders from Bargaintown ten miles away. Sweet potatoes are a year-round problem, less because of their cultivation than their storage. The coop built a storage shed, one of many in the area, but could not plant or harvest enough to make a success of the enterprise. Depending on the growing season, the condition of the potatoes at harvest, and the weather during the winter storage season, one can expect to lose from 10 to 50 percent of one's sweet potatoes before market. Storing the potatoes simply was not worth the coop's efforts, and the shed was abandoned in the sixties. The Atlantic Electric Company bought it for use as a warehouse.

Caught in the same cost-price squeeze as other farmers, Grunow finally turned to white potatoes, a crop that can be planted, cultivated, and harvested mechanically. His capital investment is heavy because harvesters cost more than $20,000 and tractors more than $30,000. Maintenance, even though he does much of it himself, is not cheap, and fertilizer and pesticide prices are exorbitant. He has too much invested in potato farming to do anything else, but his worst fear is that in his section of the township, farming may no longer be viable after he quits. Five-acre residential plots have sprung up around him, and land prices have risen sharply since casinos were built twelve miles away.

Grunow is a long-time member of his town's planning board and makes sure farm interests are represented when planning and zoning decisions are made. Galloway has lost surprisingly little farmland since 1940, although it was never a major agricultural center like Hammonton. In 1940, 11 percent of the town's 90 square miles was in agriculture, and in 1979 agricultural land was still 9 percent of the total. Developed land had risen from 3 to 17 percent of the town's areas, but almost all suburban development occurred in forest, not farming sections. Now, however, it appears that

Grunow's generation will be the last to farm major sections of the town. It is possible to preserve Galloway's farmland by zoning and regulations, but is it possible to preserve its farmers?

And what of the people still in Egg Harbor City, especially the farm laborers and industrial workers who were the foundation of Egg Harbor's wealthier days? The blacks still form a substantial minority of 11 percent in Egg Harbor City.

GARDENS OF THE BLACKS

About a mile south of Egg Harbor City stand a dozen neat houses across the road from the Atlantic County 4-H Center. This is a concentrated section of a larger black community which was settled in the 1920s. Marie Wynn, now almost seventy, still looking sixty, lives in the green house with her son, daughter, and son-in-law. "I never would like to live alone," she says. "Some old people like it alone. Not me." She has been with her family in Egg Harbor since 1927.

Before Mrs. Wynn settled in the area, she bounced around the East Coast as her father, H. L. Byrd, moved from opportunity to opportunity. H. L. Byrd dominates family memories, and, although he died in 1961, the impact of his high energy, his ambitions, and his Southern Baptist strictures still lives. Byrd, whose roots probably went deep into Tidewater Virginia, began his family in that region where he was a sharecropper and carpenter. Throughout his extraordinarily complex life, he never lost his first loves—carpentry and gardening. After Marie was born, H. L. began moving around to look for the best place for his family and his own security. By the time she was nine years old, Marie had moved at least four times, first to North Carolina, then South Carolina, then to Coatesville, Pennsylvania, and finally to Buffalo, New York. Like hundreds of thousands of other poor, rural people in the first two decades of the twentieth century, H. L. Byrd was forced off the land into steel mills. And like many of those people, he held onto his dream of returning ultimately to the land.

In Coatesville and Buffalo, H. L. worked in the steel mills; he stayed in Buffalo from 1921 to 1927 by which time he had saved enough money to buy one or two acres of land in some rural area. After reading an advertisement in a Buffalo newspaper about the attractions of Egg Harbor City—its mild climate, inexpensive land, fruitful fields, community spirit, and modern transportation—he bought about two acres and moved his family to Egg Harbor City, south of the railroad tracks. The town was and still is racially segregated regardless of the democratic layout endorsed by the founding fathers. At that time there were no more than a dozen black families living permanently in South Egg Harbor, although migrant black farm workers either from the south or from Philadelphia were common in the 1920s. As Marie, then fifteen, remembers, her father liked rural South Jersey much more than did the children, who missed their friends and the excitement of the city. Further, H. L. appears to be one of the few who liked Egg Harbor enough to want to stay. "A lot of people," said Marie, "bought land like a pig in a poke." They came out from the city, did not like what they saw, and left without turning over a furrow. Some even bought swampland, sight unseen.

Two years later, because he needed more capital and because some in the family missed the city, H. L. moved back to Buffalo and the steel mill. That was in 1929, and when the Depression hit, he was out of a job, but still had his land. Back he went to Egg Harbor, where he remained the last thirty-one years of his life.

The Byrd family got through the Depression like most other people in the Pine

Barrens, with plenty of food and little money. H. L. had his land, his garden, his carpentry, and thirty years' experience as a jack-of-all trades, so he got along. He built his own house and helped build others, and he did odd jobs in the community; he worked on a few WPA projects; he formed a car pool with a half-dozen other men during the late summers and falls to commute to Port Norris to shuck oysters; he picked blueberries in summer and cranberries in fall; and he worked his garden, some of the produce from which he sold locally or gave to neighbors. With the end of the Depression, wartime shipbuilding came to Camden, and H. L. worked during the forties at the Camden shipyards while he continued his myriad other tasks.

H. L.'s farm was about two acres, three-quarters of which were planted to corn that went to feed his hogs, one of which was so big that people from around the region came to the farm just to see its size. The rest of the garden produced the many crops typical of South Jersey: sweet and white potatoes, beans, peas, salad and other greens, squash, tomatoes, onions, various root crops, and, as Marie remembers, enormous winter beets. His crop diversity served two functions. First, it provided a well-balanced diet, and second, it ensured that, regardless of seasonal weather fluctuations, one or another crop would succeed.

Besides his garden and carpentry, H. L. loved his community and became a deacon in the Shiloh Baptist Church, the hub of social activities in South Egg Harbor. By the time he died, H. L. Byrd had led a full life, but, like Will Grunow, he left no one behind to tend his garden.

H. L.'s daughter, Marie, had married Earnest Wynn during the Depression. Earnest came from Pittsburgh to Egg Harbor City with two other boys when they were about twenty years old after they too had seen Egg Harbor advertised in the newspapers. Young Earnest, a city boy, came not for the land and gardens, but for the job possibilities in one of the small factories that still existed in the town. Meanwhile, Marie worked various jobs herself in the factories. The Byrd / Wynn families mutually supported each other during the Depression, but, as World War II arrived, the Wynns moved to Philadelphia where Earnest was a machinist during the decade of the forties and Marie worked on and off in factories to supplement their income. They returned often to Egg Harbor, Marie with the intent of raising her children "in a decent place, not in the city."

It was the basic politeness and neighborliness of Egg Harbor that held the community together and still makes it appealing to many young people. The Wynns returned to stay in Egg Harbor in the early fifties, and, although Earnest died in 1957, Marie continued her factory work. She remained a solid church member and one of the pillars of the Ladies' Club of South Egg Harbor.

The gardens, however, are gone, and Marie Wynn's children have no interest in working the soil. While rural black enclaves still exist in pockets in the Pines, young blacks, like most other young people, have opted for the mainstream. Marie's son has graduated from the local college, and her daughter teaches in the school system of neighboring Mullica Township. For the Wynn and Grunow children there will be no more gardening or farming.

PINEY GARDENS

Ten miles north of the Wynns and Grunows are the people of the woods, among them the Percy Miller family. Seventy-year-old Percy, his wife, Mary, and his son Jack live in the turn-of-the-century white farmhouse in what is called Tylertown, a relic of a larger community where lived several other families like the Adams and Gerbers, long since gone. Tylertown is typical of what is left of many small Piney places throughout the region and is equidistant between Batsto and Bulltown, an old cranberry bog, now a recreational lake. Bulltown (photo 15) is across the highway from Herman City, the aborted glass factory. The nearest real village is Green Bank, the old river community four miles southeast of Tylertown.

©*James F. Gandy, Jr.*

PHOTO I5. *Charlie from Bulltown*

Like many Piney families, the Millers live as much by the earth as by the forest; indeed, most of their food comes from their garden, which is a smaller version of H. L. Byrd's. Since Percy is now too infirm to work his plot, Jack and another son, who lives nearby, do most of the jobs. As it now exists, the garden is a one-acre plot across the macadam road from the farmhouse, and it produces, as it has for more than two generations, the beans, tomatoes, potatoes, squash, greens, root crops, and corn that have been the bulk of the Miller family's diet. Fresh meat often comes from the woods, but, in the past, most of it came from the pigs that were raised in the spring and slaughtered in winter. In addition, salt pork was available from the general store for six cents a pound.

Most Americans are used to stories of exploitative farmers and woodsmen—pioneer types—who found rich soil, mined it by extensive and wasteful practices, and moved on to new ground. Not so in the Pine Barrens, where soils were not naturally fertile. When Percy's ancestors settled Tylertown, they found infertile, acid soils that required manure and labor to make them productive. Settlers who had to put a great deal of effort and organic matter into a plot were not likely to give it up easily and move to inferior land. In the Pines the investment was what people put into the earth, not the earth itself.

The Miller garden has seen many different manures in its time. When there were pigs, chickens, horses (never more than two), and a milk cow (usually one little Jersey), domestic animals provided manure. When nearby salt-hay farms abounded, there was good, cheap mulch, and when the river trade brought clams, oysters, and mussels up the Mullica, shells provided minerals and basic elements to sweeten the soil and make it less acid. Prior to 1960, when the menhaden factory in Great Bay still worked, cheap fish meal was available, and before that the Millers could use alewives, the local anadromous herring. The March or April alewife run stopped at the Batsto dam, where the small fish would congregate to breed by the thousands, there to be scooped up and smoked, eaten fresh, pickled, or plowed under the soil.

Most Piney gardeners now use some chemical fertilizers, but as one Piney said, "Fertilizer? Fertilizer's all right if, like some guy needs a shot o' booze, you know, to pep him up, that's what fertilizer is. But you got to put something in the ground to hold it, like organics. We'd put manure down the row and throw two furrows to it and plant some sweet potatoes. You'd get twenty #1s to the hill."

Percy kept his plants free from pests by keeping the soil healthy, thus in turn giving his garden a chance to ward off the depredations of fungi and insects. He also rotated crops, although in no particular order; he simply made sure that tomatoes or beans were not planted in the same place every year. Until insecticides were cheap and readily available, he would pick off tomato worms and potato and bean beetles by hand or dig out with a penknife the large grubs that bore into stems of squash. Percy also let birds help with insects. Although an inveterate hunter, he never killed quail. "In the summertime," he once said, "I'd watch them quail when I used to have potatoes out there. Ya know, them quail go right down that row there and pick the bugs off'n the potatoes, and ya didn't have to bug 'em yourself."

When Percy was growing up in the 1920s, his garden was simply one of many in the community. Most of the gardens are now gone, grown up in forest. Before we leave him, we should give one last look at the way his garden and old field surround him.

We have seen the acre of garden just across the road. But there is also the open field he keeps cut. It stretches about three hundred feet around the perimeter of the house and stops at the forest line. In it he used to plant corn when he had animals to feed, but now he just keeps it cut. To what purpose? "One spring day," said Percy, "sometime in the 1930s, I don't remember which year, a fire come down on us. Probably started all the way up around Chatsworth. I still had some cornstalks in the field, so when the fire come close enough, I set 'em ablaze for a backfire, and the fire went right around the house. Left us alone, just like a hole in a doughnut. Fire went all the way down to the Bay. Yeah, left us alone. That's a double-duty garden."

Chapter 4

FIRE

The fire-prone forests of the Pine Barrens give the region its character (maps 7a, 7b). In the waters lie the richest resources and historic transportation routes, and the earth yields a surprisingly wide array of food. Still, it is the forests that serve as the backbone of the Pines.

To Pine Barrens residents, the woods are for living in, for hunting and walking through, and for seeing the nostalgic places of one's youth (photo 16). As Janice Sherwood said, ''A Piney is just a little deeper in the woods than you are.'' The woods are where eighteenth-century pirates lived and through which revolutionary contraband traveled. The forests are filled with lost towns, such as the spot where Mr. Ong threw his hat in the air, now called Ong's Hat. The trees of the forest are cut for boatbuilding, dwellings, furniture, banjos, pulp, charcoal, and wood stoves. The iron and glass industries were dependent on wood, as is boatbuilding, which, along with carving, remains one of the premier crafts of the Pines.

Above all, the pine forests are important to outsiders—scientists, environmentalists, hunters, hikers, and canoeists. These groups together with a few well-placed insiders, who prize the unique ecology and contiguous forestland, formed the most important lobby to pass Pinelands legislation. Few in number but politically powerful, they wrote the books and led the fight for Pinelands preservation. The most important

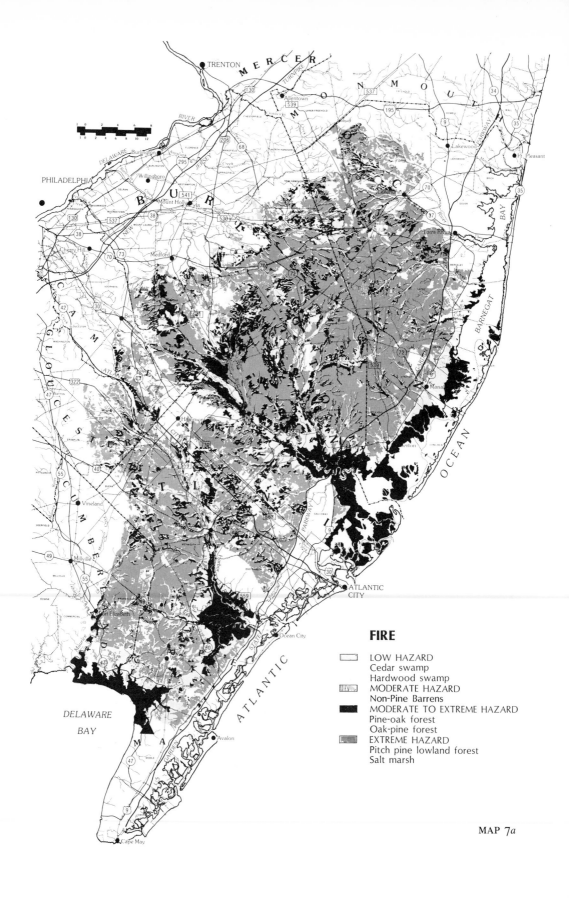

FIRE

LOW HAZARD
Cedar swamp
Hardwood swamp

MODERATE HAZARD
Non-Pine Barrens

MODERATE TO EXTREME HAZARD
Pine-oak forest
Oak-pine forest

EXTREME HAZARD
Pitch pine lowland forest
Salt marsh

MAP 7*a*

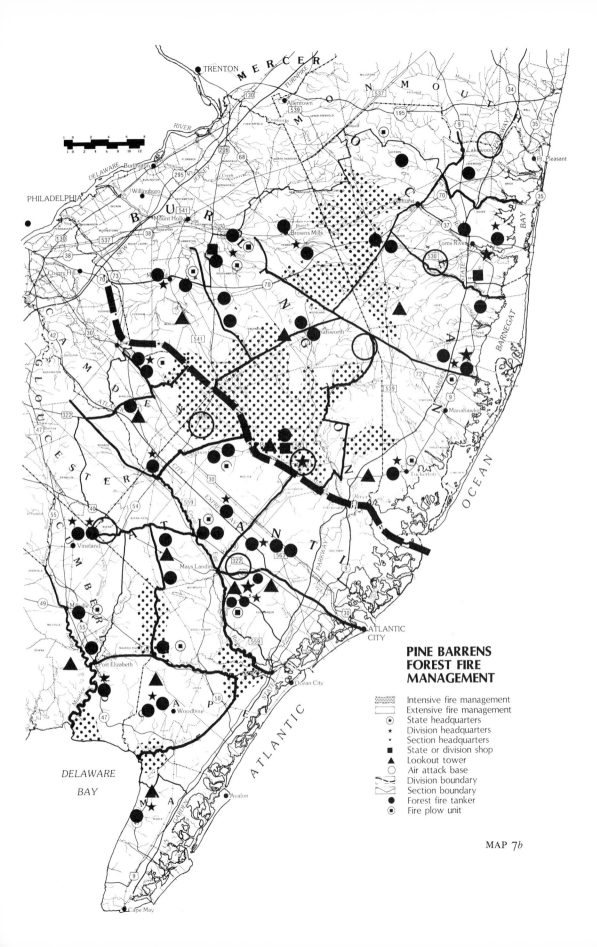

PINE BARRENS
FOREST FIRE
MANAGEMENT

Intensive fire management	
Extensive fire management	
State headquarters	
Division headquarters	
Section headquarters	
State or division shop	
Lookout tower	
Air attack base	
Division boundary	
Section boundary	
Forest fire tanker	
Fire plow unit	

MAP 7*b*

PHOTO 16.   *Sand Road near Weymouth Furnace*

©*James F. Gandy, Jr.*

of them were Jack McCormick, who focused his attention on the scientific community and the National Park Service, and John McPhee, who was a great publicist and confidant of Governor Brendan Byrne. As McCormick (1979, 229) wrote in the last article he published before he died in 1978, ''The modern vegetation of the Pine Barrens is unique in the world. As a natural open space and recreation resource, the region also is of overwhelming significance to the state, to the megalopolitan region, and to the nation.''

In this chapter we look closely at the pine woods—the forces, chiefly fire, that created them and the people who use and shape them. First we will describe what the forests look like ecologically; then we will look at their history, the defunct rural industries that depended on them, and the impact of fire; and finally we will describe their present use by insiders and outsiders—for lumbering, plant collecting, hunting, scientific study, and recreation.

Forest Composition and Succession

The kinds of trees that compose the forests of the Pine Barrens are, as in any other forest, dictated by four factors: climate, soil (and the amount of water in the soil), the age of the forest (its successional stage), and disturbances in the forest such as fire, cutting, or clearing for agriculture. The three elements that give the Pines their distinctive appearance are the swampy lowlands, the sandy soils, and repeated disturbance.

Because the region's soils are so sandy, acidic, and either swampy or droughty, the species of trees that can live in such a place are limited to those that can tolerate acid, drought, or standing water. Such conditions preclude many trees common to most mid-Atlantic regions such as tulip poplar, sweetgum, sycamore, hickory, willow, and white pine.

Nineteenth-century botanists divided Pinelands forests into several categories, and the modern classification rests on the work of Silas Little (1979a; 1979b) and the late Jack McCormick (1970). The classification is based on the kinds of disturbances that occurred and the presence or absence of water in the soil. On wet soils (lowlands) one could expect to find Atlantic white cedar and/or a variety of swamp hardwoods, predominantly red maple, black gum, and magnolia (sweetbay). On transitional sites are often found pure stands of pitch pines; and upland soils are occupied by pitch or, less often, short-leaf pine and a wide variety of oaks, depending on the richness of soil. On poor soils one most likely would find scrub, blackjack, and post oak with some black and scarlet mixed in, while on richer sites, black, scarlet, white, and chestnut oaks.

The species composition of a forest also depends on its stage of ecological succession through time, that is, how long it has been left undisturbed by fire, cutting, or wind damage. Anyone who has watched the history of a woodland knows that, after a forest fire, oaks and maples do not immediately reappear in place of the old trees. Instead, a forest goes through stages by which trees that cannot tolerate shade (pines) first establish themselves, later to be followed by trees that do tolerate shade (oaks). Table 2 is a simplified diagram of forest succession in the Pine Barrens.

Vegetation succession is only theoretical, because the forests are so often disturbed. In some forests, such as the redwoods, succession is "fixed" because the climate is benign and catastrophes infrequent; such forests may live thousands of years with little change in species composition. As Little (1979b, 312–13) wrote, however, this is not so in the Pines, whose

> vegetation has been shaped in large part by extensive wildfires and heavy cuttings, and locally by other disturbances. Abandoned upland fields progress, in order of dominance, from grass to pine to oak, but the shrub layer develops slowly. Abandoned swamp sites invaded by shrubs, Atlantic white cedar, or hardwoods, also tend toward hardwoods. Cutting may hasten succession toward hardwoods in

TABLE 2. *Forest Succession in the New Jersey Pine Barrens*

Amount of Water on Soil Surface	Successional Stage		
	Youth (0-100 yrs.)	Middle Age (100-200 yrs.)	Maturity (more than 200 yrs.)
WET (Standing water in March)	Cedar Seedlings ⟶	Cedar Swamp ⟶	Cedar and/or Hardwoods
	Hardwood Sprouts ⟶ and/or Seedlings	Hardwoods ⟶	Hardwoods
	Mixed Cedars ⟶ and Hardwoods	Hardwoods ⟶	Hardwoods
INTERMEDIATE (Moist Soil in March)	Pitch Pine ⟶	Pine/Oak ⟶ (Drier Sites)	Oak
		Pine/Hardwood ⟶ (Wetter sites)	Hardwoods
DRY (Dry Soil in March)	Pine/Oak ⟶	Oak with some Pine ⟶	Oak

SOURCE: *Sinton (1977). Reprinted with permission from* Environmental Review.

both uplands and lowlands. Successional rates vary enormously: whereas an upland field may become a thicket of pine saplings in 15 years, leatherleaf shrubs may dominate a lowland spot for 50 years before trees begin to dominate.

Different fire frequencies and intensities interrupt succession, thus accounting for most of the current variations in forest composition, and largely obscuring the effects of soil differences. In swamps, if fire consumes enough of the organic soil, quaking bogs, meadows, or leatherleaf areas may be favored. On uplands and lowlands, many pines are sprouts from old root crowns, or have recovered from the last fire by trunk sprouts. Fire kills oak stems more readily than pines, but most oaks sprout. Periodic wildfires at possibly forty-year intervals have produced oak-pine mixtures over extensive areas of upland, while more frequent fires have created mixtures of pitch pine and shrub oaks, and most frequent fires created the Pine Plains.

A bewildering variety of things can happen to a place in the Pine Barrens after a disturbance. In fact, disturbances are responsible for the patchwork pattern of the forests, which, to untrained eyes, appear uniform. Disturbance is also responsible for the continued presence of many of the rare and threatened species of plants and animals in the region.

Most unique plant and animal species in the Pines are adapted to withstand frequent disturbances, especially fire. These species are found on sites where tall trees do not dominate the landscape—around bogs and on the more open uplands. In swampy areas and at the edge of wetland forests thrive the curly grass fern and yellow-fringed, crested yellow, and southern yellow orchids along with two dozen lesser-known grasses, rushes, asphodels, bladderworts, and ludwigias. On the open uplands, fields, and roadsides grow the Pine Barrens gentian, Pickerings morning glory, broom crowberry, slender-leafed goldenrod, and slender rattlesnake root. None of these can survive in late successional forests, nor can the Pine Barrens tree frog. Even the rare pine and corn snakes and timber rattlesnakes fare best in old fields and young forests where abandoned dwellings provide shelter and denning places.

The most extraordinary ecosystem in the Pine Barrens—the Plains, also known as the Pygmy Pines or Short Hills—are adapted to, even dependent on, frequent fire (photo 17). A person can stand on a knoll in the Short Hills and look over the tops of the forest trees for miles in all directions. This is one of the few true pygmy forests in North America, except for those in subalpine and alpine zones. Small patches of similar pine/oak forests also occur near Albany, New York (the Pine Bush), and on Long Island. Another pygmy forest occurs in a most ironic place, Mendocino County, California, where some of the tallest redwoods grow (Good, Good, and Andresen 1979). Frequent fires have probably ravaged the Plains for thousands of years, but no one can be sure. The general boundaries seem to have remained stable since the seventeenth century, although it was not until the early nineteenth century that the first citation in literature referred to what was then called the "Grouse Plains" (after the now-extinct heath hen) (Good, Good, and Andresen 1979). The first land surveys were taken in the 1830s; since then, Plains boundaries have not changed.

PHOTO 17. *Pygmy Pines and Land Scam Billboard, 1934. The billboard reads: For information for these unusual acres—Address, Lloyd Conover, Realtor, Clinton, N.J. Note the natural desirable landscape—Take in the superior beauty of these rolling acres.*

Under its strict fire regime, the dominant tree, pitch pine, responds by sending its roots as deeply and widely as possible. More than most trees, pygmy pitch pines lead underground lives. Instead of sending a single trunk skyward and a taproot downward, fire, and perhaps genetic messages, forces them to form multiple spreading roots and as many as two hundred spindly sprouts. To dig in the sandy Plains soils is to cut into countless tangled roots; following them is a tedious job, rather like a careful archaeological dig. The root crowns are as old as one hundred years and from them grow sprouts, none older than twenty-two years (Good, Good, and Andresen, 1979). Apparently, the old roots tend to lose their vigor over time, and this is probably one reason for dwarfism in the Plains.

Scientists have given many possible causes for the Plains, such as infertile soils, hardpan or ironstone not far below the surface, deep winter freezes, high winds, and aluminum toxicity. Most likely the Plains are a result of physiological adaptation to fire, such as the old root/young sprout phenomenon, and genetic differences between populations in the rest of the Pines. The strains of pitch pine in the Plains, for example, have exclusively serotinous cones that open and throw seed only after the heat of a fire. These pines also tend to exhibit a spreading rather than a vertical growth habit regardless of fire, and they produce seed at an earlier age than other pitch-pine varieties (Forman and Boerner 1981).

Unlike the great forests of the Appalachians, the Adirondacks, and the western mountains, those of the Pine Barrens have acquired their pattern and uniqueness from frequent disturbance. This is not a forest wilderness, but a woodland long manipulated by people. Disturbances themselves have become the balancing mechanism whereby the patchwork patterns and unique species can thrive. The insiders work constantly with fire and cutting to create patterns and habitats, while the outsiders enjoy the scientific and recreational opportunities that such work has created. The only historic constant in these forests has been change.

The Piney Woods

THE PREHISTORIC WOODS

About 12,000 years ago the last glacier left New Jersey. The glaciers had probably come as far as the Raritan River just north of the Pine Barrens, and glacial rivers and streams poured out the sands and gravels that now lie on top of many parts of the northern Pines. A dense spruce, fir, birch, and alder forest lay on the soil, and the landscape probably looked much like that in Maine today. Aboriginal people had already arrived and were living off the caribou, mastodons, and mammoths, which were the dominant mammals of the taiga wherever it then existed in the northern hemisphere.

The climate warmed quickly over the next 2,000 years, and from about 10,000 years ago until the first Europeans came, Pinelands forests appear to have been remarkably stable. Scientists studying pollen grains have carefully followed the history of these forests and have found that pine, chestnut, oak, and hickory dominated upland sites, while holly, cedar, black gum, birch, and red maple were common on wet sites (Hartzog 1982).

The extraordinary stability of the forest suggests that the aboriginal groups who lived in the area had achieved a reasonable balance with the nonhuman system despite climactic fluctuations and a change in game animals from caribou and pachyderms to deer and passenger pigeons. The forests had more diverse vegetation then than now, although the patchwork pattern of alternating open- and closed-canopy areas probably was absent because Indians had no need to burn extensively, and their population was scattered in small villages.

Eighteenth- and nineteenth-century ethnographic accounts drew conclusions that seemed appropriate at the time. Travelers and scientists saw the Pines as barren and inhospitable to white and Indian alike; Indians were, therefore, unlikely to have lived there. And, since early writers were predisposed to search for large fortified villages, such as those of the Iroquois in 1700, they were not attuned to look at signs of small permanent and semipermanent settlements. Until recently the general picture of Indians in South Jersey was as follows:

> The great mass of villages and cemeteries . . . are along the Delaware River and its tributaries. . . . It will be observed that the interior is practically devoid of sites, except on the headwaters of the more important creeks and rivers. This lends support to the tradition that the sandy interior of South Jersey was more of a hunting ground than anything else. (Skinner and Schrabisch 1913; 16, 41)

And more recently:

> A large portion of the state, especially the Pine Barrens and the southern swampy areas, were not attractive to an aboriginal or even to an early historic people. (Cross 1965, 4)

In the 1970s a group of archaeologists challenged prevailing theory. They began to look harder in the Pines for sites, assuming that most known Indian complexes were in agricultural areas where soils had been tilled and plowing had made it easy to find artifacts. The Pine Barrens, they knew, was the only area along the northeast and mid-Atlantic coast where soils were not extensively cultivated and where one might find some of the few remaining undisturbed prehistoric complexes. In amassing old data and digging likely sites in the Pines, archaeologists discovered more than a thousand sites, some 10,000 years old, some from the days of European contact, and others used by different groups over thousands of years. Like the European settlements, those of Indians were concentrated in coastal and riverine areas where resources were richest. A clear picture is just now beginning to emerge of extensive Indian exploitation of the Pine Barrens. The oak uplands, while they could not support large populations, surely had sufficient nut crops and scattered fertile soils to hold smaller permanent groups, and the large number of artifacts in many different areas gives this theory credence.

By the time of European contact in the mid-seventeenth century, the Indians who then occupied the Pines—the Lenni-Lenape, a group of Delawares—had a mixed foraging and farming economy. Alan Mounier (1982, 117), a local archaeologist, wrote:

The pattern that emerges is one of efficient adaptation to an essentially modern environment with few resources being overlooked. The prevailing model of early historic Delaware settlement suggests that the population, while tied to a given territory, was committed to an exploitative regime which required flexibility and a considerable degree of mobility.

THE HISTORIC WOODS

The first Europeans to arrive were the Swedes on Delaware Bay in 1638, followed by a series of Long Island and Connecticut Yankee settlements on the coast from 1640 on. The Lenni-Lenape were quickly killed by smallpox and other diseases, or moved west, so by 1750 only a few hundred Indians remained in the whole state (Cross 1965). The few remaining were put on a reservation in the Pine Barrens called Brotherton, now Indian Mills, and finally left in 1801 to join a group near Oneida Lake, New York.

As in the rest of colonial America, settlement patterns moved along rivers, and one finds the first recorded sawmill in the Mullica Basin in 1730 (Pierce 1957). The bulk of timber and wood products went to water-based industries, and the larger oaks and pines were all probably cut within the first hundred years of settlement and either sent to England or used for local shipbuilding and house construction. By the middle of the eighteenth century, sawmills were scattered on all the rivers throughout the Pines, and around all these millstreams appeared small settlements, the basic outlines of which one can still see in Browns Mills, Wading River, Batsto, Tuckahoe, Dennisville, Port Elizabeth, and dozens of other towns.

In the middle of the eighteenth century, industry came to the middle of the Barrens. Beginning in 1765 with the iron furnace at Batsto, rural industries appeared in every section. Between the 1760s and 1850s some two dozen furnaces worked, along with a smaller number of forges and slitting mills. In 1800 a series of twenty glass factories appeared, many of which closed by 1870, some of which lasted to 1930, and a few of which were transformed into the modern glass industry around Millville. In addition, there were several paper mills and cotton factories, the latter producing cloth from rags, not raw cotton.

The major natural resource on which rural industry rested was wood, and because of this, people continually cut the woodlands, from the smallest pitch pine to the largest cedar. The enormous demand for wood coupled with uncontrollable forest fires drastically changed the landscape of the region from its prehistoric form.

A medium-sized iron furnace, such as the Richardses' Batsto works, required about 6,000 cords of wood per year for the 12,000 bushels of charcoal used to fire its furnaces. Forests in the Pine Barrens generally produce only one-third to one-half a cord of wood per acre per year (Pierson 1979). A conservative estimate is that the five furnaces in the Batsto area, then, would have required 40,000 acres of woodland annually for their production. But in fact more wood was needed because in that same area were three forges, four glasshouses, a paper mill, and a cotton mill. Furthermore, oak had to be cut for domestic fuel and cedars for boatbuilding and shingle production. Coppice, or scrubby oak, was cut for hoop-poles, for barrels, for pole wood, and for basketmaking. Most economic activities in the Pines depended on wood (and still do), and hardly a man was born in the Pines who, at one time or another, did not cut wood or fell trees.

Table 3 explains the major uses of the three most economically important trees in the Pine Barrens.

The white cedar of the wetlands was and still is by far the most valuable tree in the Barrens, as much because of its scarcity as its irreplaceable qualities of light weight and resistance to weather and disease. It is in high demand for roofing, fence posts, and boats. Oak is most valuable as a mature tree for construction and in small pieces for heating fuel. Pitch pine is the most useful because it has been essential to industry. Small trees are mostly cut for charcoal, but some large pines are used for construction, poles, and railroad ties as well as for newspaper pulp.

The insatiable demand for wood from the Pines throughout the nineteenth century required that the whole region be cut several times over. Each year the woodcutters and colliers combed their areas for new sources of wood and cut everything one inch in diameter and up. They moved to the next wood lot, leaving slash and sand behind them, and sometimes even purposely fired over the site so they could purchase it cheaply and harvest it fifteen years later. Further, wildfires claimed tens of thousands of acres of forest every decade. Reports from the *New Jersey State Geologist* in the late nineteenth century are filled with descriptions of fires that burned ten, twenty, or thirty thousand acres at a time. A hundred years ago much of the Pinelands looked barren indeed—a forest devastated by axmen and wildfires.

TABLE 3. *Value of Forest Products in the New Jersey Pine Barrens in the Nineteenth Century*

LEGEND: High Value ■   Medium Value ●   Low Value †

	1800	1850	1900
A. Cedar			
1. Shakes for Roofing	■	■	■
2. Poles and Posts	■	■	■
3. Planks for Boats	■	■	■
B. Pine			
1. Poles and Masts (Mature Pine)	■	■	■
2. Boards and Planks (Mature Pine)	■	■	■
3. Railroad Ties (Mature Pine)	†	●	■
4. Charcoal	■	■	●
5. Domestic Fuel	●	●	●
6. Tar and Turpentine	●	●	†
7. Pulp	†	†	■
C. Oak			
1. Boards and Planks	■	●	†
2. Domestic Fuel	■	■	●
3. Charcoal	●	●	†
4. Barrels and Baskets	●	●	†

SOURCE: *Sinton (1977). Reprinted with permission from* Environmental Review.

Did nineteenth-century people understand that they were destroying their forests? Perhaps we need to understand what destruction and creation meant to them and to explore what was economically important to those people rather than aesthetically important to us. At what point in the life of a forest—when in terms of ecological succession—could the rural industrialists expect to get the species of trees they needed to support their economy? Put most simply, how could people in the nineteenth century get the most cedar and pine in the shortest time possible?

Had the people of the Pines been ardent conservationists a hundred years ago and protected their forests from fire and cutting, they would have produced precisely the kind of woodland least useful to their economy, since climax conditions result in oak and mixed swamp-hardwood stands. People need pine and cedar, which generally occur early in succession and, in fact, cannot regenerate extensive stands unless they *are* disturbed. By cutting forests and starting fires intentionally, the people of the Pines were, in ecological terms, keeping the forest in an early successional stage and, in economic terms, assuring themselves of the kinds of wood, especially pine, necessary to support their industries.

In 1850, after three generations of heavy cutting and burning, the Pines were certainly barren. Photographs from the turn of the century show desolate backgrounds, barren land, and slash piles. But the regenerative powers of even this poor soil were sufficient to supply nineteenth-century industrial needs. Out of the ashes and stumps and roots grew the pines that made the wood and charcoal that ran the furnaces, forges, and glasshouses. Every twenty years the wood choppers and colliers gathered their harvest, and the cycle of the phoenix continued until the industries died (Sinton 1977).

This dramatic dynamic is the background of Jack Cervetto's philosophy of forest use. Jack, who lives in Warren Grove in southern Ocean County, believes "the forest needs a good fire every so often." Without formal courses in ecology or silviculture, Jack knows the history of every area in his district. He knows firsthand about the effect of fire on acorns and deer and about the regeneration of cedar and pine. Jack straddles the older, more exploitative era of use and the modern era of fire control. Today's Pinelands forest is a balance between these forces. The massive disturbances of earlier times, the dependence on the forest for basic resources, and the fire fighting of the last forty years provide today's users with a choice of landscapes.

Fire Ecologists

FIRE IN THE PINES

Twenty miles northwest of Tylertown, across the woods of Wharton State Forest and Sim Place bog, between the dwarf forests of the East and West Plains and the Garden State Parkway, sits Jack Cervetto's house, in Warren Grove. Jack, like many other Pineys, was not born in the area. He was born in 1908 in Clifton, just north of Newark. When he was a boy, a peddler came on his rounds past his parents' house. The peddler needed a horse and was willing to trade his deed of fifty-six acres of Warren Grove for a horse, so Jack's father made the deal. During World War I Jack's father stayed in Clifton and sent the family down to the little house he had built in the Pines. Jack's father never did like Warren Grove much and moved the family back to Clifton after the war. At the age of twenty-one, with five dollars in his pocket, Jack left North

Jersey and never returned. He married into one of the old Warren Grove families and has spent most of the rest of his life cutting wood, fighting fires, tending his two-acre garden, and raising his family.

Jack has lived the past fifty-two of his seventy-three years in the little village tucked away in this most fire-prone area of the Pine Barrens. We call him a fire ecologist because he works with and fights fires; Jack knows how to adapt his life and his town to fire and can no longer count the fires he has fought. He remembers only the big ones that he can still see in his mind's eye—especially the fire of 1936.

Fire conditions were as bad in '36 as they had ever been. A dry spring followed a dry fall and winter. People in the Pines have become accustomed to thinking about fire during "spring season" between the last week in March and the middle of June, when winds are shifty and humidity is at its lowest levels of the year both in the forest and on the forest floor. Jack Cervetto was getting edgy; the weather continued dry and windy, and he knew that the duff and leaf litter in the forest had built up to the point where it provided a lot of potential fuel for fires.

The fire of '36 started somewhere in the middle of the Pine Barrens, some say near Chatsworth, some say even further north. The fire was carried on a heavy north-northwest wind and quickly took on its own life as it began gobbling up duff, tree trunks, branches, even up to the crowns. It came down out of Chatsworth and gathered speed when it hit the Short Hills of the West Plains, where the dwarf trees provided fuel and ventilation.

Some fires just hiss along the ground and some move with a roaring hiss, shooting out along the forest floor, charring trunks. This fire came down with the roar of freight trains and cannons. It ran on the ground and leaped from crown to crown, exploding the tops of pines full of dry pitch, throwing them hundreds of yards downwind, disdaining any attempt of the fire fighters to stop the fire by digging trenches or burning the forest floor in front of the main blaze in order to rob the fire of its fuel.

When the fire got to the Short Hills, Jack knew they could not stop it. Every time he and the fire fighters dug a trench or doused a smoldering blaze or started a backfire, the wind would shift, and the fire would shoot off in another direction as the fighters scrambled to stay clear of its head. The fire was so massive and winds so shifting, they often could not find its head. Rolling fires came out of it—strips of thirty-foot-wide gaseous flame driven by side winds and their own energy; from the air the heads of the fire would have looked like they had many tongues. The rolling fires looked like balls that rolled out of the head, stripping the vegetation clean and leaving the rest standing until the wind shifted again, and then what was left went up in smoke.

The fire burned everything everywhere. "That fire," said Jack, "when it come down from the Short Hills, it come down there so hard that when it hit the cedar swamp, it just burned the tops off the cedar. Burned off fifty acres of my cedar." In its wake the fire left a charred forest and one Warren Grove warden burned and crippled for life. The only thing that stopped the fire was the water of the bay ten miles east of the village.

Jack and his neighbors, however, saved their houses by backfiring. Jack set his backfire in his cornfield, which covered a one-hundred-yard perimeter around his house. As the fire came down on the village, the backfires began to burn away the fuel on the ground that would have fed the head fires. The head fires created an enormous draft, sucking air into themselves to help digest the thousands of acres of woods they were eating. As they bore down on the houses, they sucked in the backfires that were creating their own drafts, and when the two fires hit, they slammed into each other with the roar of two careening locomotives. Flames and smoke shot high in the air. Burning charcoal was thrown onto the house roofs, there to be doused by the women and young children who had marshaled themselves for the occasion. One neighbor's house began burning seriously, and the others ran to help put it out.

"Well," says Jack, "anyone that's lived in the woods any length of time, they prepare themselves for forest fires. That's our public enemy #1. We know they'll be here. It just goes with this type of living."

FIRE-ADAPTED VILLAGE

Jack Cervetto has spent forty years as a district fire warden around Warren Grove, in Stafford and Little Egg Harbor townships. The small woodland village is the locus of the fire district, and Jack, as the man on the spot, knows the area better than the state fire boss from the region or the people from Trenton. He does not command the full body of fire fighters when they convene to fight a large fire, but, because of his local knowledge, he can often predict how the fire will spread and where it might be stopped. Jack is a technician who has learned through doing that the best way is to fight fire with fire. For every blaze he has fought Jack has backfired a lot of ground. As a burner of ground and fighter of fires, he is as much a part of the region's fire history as are the wildfires themselves.

During fire season, the people of Warren Grove often have fire on their minds. Over time they have evolved a village form that will hinder the spread of fire, aid in fire suppression, and facilitate a safe and quick evacuation. The scattered, as contrasted to clustered, development of houses is not wasteful of space but is protection against fire. The wide swaths of fields around each house effectively impede fires, and residents supplement the clearings with thinning of forests and burning of forest litter at the edges of their fields. While the meadows and gardens act as firebreaks, the spacing of houses serves to keep flammable objects separated. Woodland villages have elongated layouts and rarely have a town center. There are screens on all chimneys to cut down hazards from sparks, and few decorative shrubs around buildings: native evergreens, like laurel, are aesthetically pleasing, but they are also excellent fuel.

In a central location stands the volunteer fire company, started to fight not house fires, but forest fires. The older men constitute the main crew of the district fire-fighting troop. Although most are retired today, they have trained their younger replacements. The dispersed layout of the village allows the fire engine and crews easy access to all homes and to the forest edges. The open nature of the settlement allows larger fire crews and machines to operate without interference and provides residents with quick and easy evacuation if a large fire cannot be controlled. Finally, the initial placement of the village east of a large swamp provides a measure of protection from fires that come from the west.

It is from the west and northwest that major fires will come, as brisk spring westerlies carry the fire from the Plains down toward the village. The scrub oak and pitch pine of the Plains are the most hazardous fuel source in the Pines, while the swamp of the Oswego headwaters is the least hazardous and will burn only under extreme drought conditions. The swamp, called Sim Place, is a Warren Grove buffer that has stopped many fires; large oaks at its marginal uplands testify to its efficacy.

Despite precautions, fire will come, and Jack and his neighbors will be ready. When he looks at the forest, he sees the entire fire history of the area laid out before him. He knows when each site was last burned because he filed reports on all the fires and, as a history buff, he has read old reports and journals on fire history. Jack founded the local historical society and has collected lists of old local names, some of which he correlated with descriptions of old fires.

Insiders like Jack, however, are not the only observers of fire in the Pines. For many decades scientists have also noticed and written about changing forest patterns and the impact of fire on those patterns. Since 1930 the number of acres burned annually in the Pines has dropped sharply, even though the number of wildfires has stabilized at about 1,100 per year over the past forty years. Due to better fire-fighting equipment and tactics, the size of most fires tends to be small, no more than several hundred acres. Large wildfires that consume 50,000 acres at a time are no longer common, although they do occur in drought periods (Forman 1979). Today the Pines are a mosaic of fire-caused patches. The smaller, more frequent fires are laid on top of the rough regional grain supplied by the larger ones. As fire disturbance declines, trees will tend to grow larger, and trees not adapted to fire will appear. Indeed, the forest composition of the region is even now slowly changing, and the fire ecologists will have to adapt themselves accordingly.

Meanwhile, Jack Cervetto continues to cut cedar and fight fires while others gather plant materials or hunt game. The Piney woods are still in constant use.

LOGGING

Jack Cervetto, in his seventies, still goes often to his swamp to cut cedar. He laughs when people say it is hard work for someone his age; he loves his swamp and likes to talk about cedar music:

> You're standin' there in the cedar bottom, there's a light breeze blowin', you got these trees rubbin' against one another. They make all kinds of music. You think it's an instrument there. Boy, what a sensation that is. Then you see a little red squirrel up there. Sometimes you just stop there for awhile. And sometimes a deer's there. He's just as curious as you are, you know, wantin' to know what you're doin' there. And you look at him. Boy, there are some nice big ones in the swamp. Yeah, yeah.
>
> I like other woods, yeah, the Plains. But if you get right in the middle of a cedar swamp with a good growth of cedar, you really think that you're the only person in the world. You're closed in there in a way that gives you the feeling there's nobody else around. And the smell. Yeah, I love that sphagnum smell. Boy, that's somethin'. I found out that's why I like workin' on moss and cedar.

Like most other cedar harvesters, Jack cuts timber on special order. Cedar stands are so small and scattered in the Pines that it is generally not feasible for large companies to work them. Some cedar is leased by absentee landowners and cut by logging operators, but most is owned by loggers themselves. Jack says the amount of wood a logger gets in a day "all depends on how strong you are and how long you can stand it.'' Cutting on order gives Jack problems because he has to do it year round. If he had a choice, he would cut only in winter, when the ground is frozen and it is easier to move equipment in and out of the swamp. In spring it is often impossible to work the swamps because the water and muck come up to waist level.

Jack's hardest job is road building. ''First you have to see where your advantage is, where you can get your equipment in and haul everything out.'' He may spend half a day figuring his advantage in this tactical struggle. Cedar likes company; the trees stand close together, shallow roots intertwined in the peat moss and organic muck, shielding one another from high winds that easily topple a tree left standing alone. Jack will find his advantage where the nearest transitional and upland soils grade into the swamp.

''Most times,'' says Jack, ''you have to work a week to make a causeway.'' The causeway, also called a crossway or a pole road or, in the West, a skid road, is a wide swath through the woods to the cedar stand. The roadbed is composed of the stems of swamp maples placed closely together so a person can walk on them, or, when Jack uses his jeep, close enough so tires will not get caught.

Next comes the work of cutting, which, for Jack, is the enjoyable part of the operation. Prior to 1960, Jack used an ax that he honed to a fine point and could bite three inches into a cedar at one cut. He could, in his prime, cut three to four cords of wood a day. He now uses two chain saws and still cuts the same amount, but with less effort. ''A good axman can do as much as a chain saw. A chain saw can go a half hour heavy, and then it has to rest. I have to count breakdown time, too.''

Jack tries to fell the trees directly onto the causeway, cuts them into eight-foot lengths, and hauls them out with a jeep and chain. Like many other one-man cedar loggers, he used to have his own sawmill, but now he just cuts poles because the demand is steady and he found milling a headache. He began his cedar-pole business in 1960, when the construction of coastal vacation houses in Ocean County increased dramatically. The following decade, developers worked their will on the salt marshes, draining and ditching them in imitation of the patterns in South Florida. The developments required an enormous quantity of cedar for pilings, docks, and bulkheads, and Jack provided much of the timber for the developments. Coastal marshes have been strictly preserved since 1970, so now Jack simply keeps up with the steady demand for poles for marking clam lots or crab pots and for refurbishing old docks.

The price of cedar has always been high. An eight-inch-diameter pole sixteen to eighteen feet long sold for about $5 in 1960 and in 1980 was worth $8. Jack no longer bothers to cut the trees into eight-foot lengths, but sells them according to the diameter of the butt at $1 an inch—$5 for a five-incher, $8 for an eight-incher, and so on. He makes reasonable money and loves the work.

Jack does not log the uplands for pine or oak, but, until the energy crisis of the seventies, few people did. Some logging companies in the Delaware Valley, on the edge of the Pines, take pitch pine regularly for newspaper pulp. State foresters often lament that the region's timber resources are sadly underutilized and badly managed because of lack of care, poor seed stock and genetic strains, and uncontrolled fire. The demand for fuel wood, however, was so low for so many years that lumbering was not economically viable. Since 1975, however, demand has jumped so quickly that fuel wood has gone from $25 a cord to more than $75 a cord. Considerable logging continues in the Pinelands; there are at least twenty-five sawmills, nine pulpwood cutters, and seven contract loggers plus an unknown number of small loggers within the Pine Barrens and on its borders (N.J.D.E.P. 1980).

GATHERING

Historically, logging is only one of the small-time economic ventures in the Pine Barrens, but today, together with the gathering of plant materials for decoration, it is the only such activity that remains economically viable. The last of the charcoal makers quit in 1975 (Gordon 1982), and the last peat-moss gatherer finished work five years earlier. While picking wild blueberries remains a family diversion, it ended with the development of cultivated berries in the 1920s. Still, a surprising number of people make a reasonable living gathering plants.

Leo and Hazel Landy gather, dry, cure, pack, and sell a wide variety of Pine Barrens plants to wholesale and retail florists (photo 18). When asked what kind of work he does, Leo says, "I'm a gatherer." Gatherers have a special place in the history of North American botany and horticulture. For every famous plant hunter who visited North America on journeys of exploration, for every famous botanist who collected and classified plants, and for every famous gardener and horticulturist who sold plants, bulbs, and materials, there had to be a Leo Landy. Rarely were these people credited for their work. Few people know the assistants of John Bartram, Peter Kalm, or J. W. Harshberger. Philadelphia was one of the centers of the development of American botany, and these unknown but officially titled "collectors" played a significant role.

A small but steady market has kept the Landys and their helpers in business. Since colonial times, Americans have decorated their homes with arrangements of dried and fresh flowers. Early English and colonial "domestic-economy" manuals carried directions on the use and preparation of common weeds. In the Victorian period, formal flower arranging became a common form of interior decoration, and almost every illustration of seasonal fashion had bouquets of dried materials in the background.

Today flower arranging continues as a serious craft, recreation, and business, but formal arrangements represent only a fraction of the interest in dried flowers. Many, who never think of competitions or flower shows, gather their own materials or buy from florists to make seasonal arrangements. Leo and Hazel Landy collect many of these materials, and quite a few are located very close to their home in Nesco, just south of the Mullica River.

Nesco, an agricultural and woodland village in Mullica Township, sits on a dry rise of tertiary deposits. To the west are the fields and orchards of Hammonton, to the east is Egg Harbor City, but around Nesco are simply fields and second-growth woodlands. Just off the dry rise are large cedar swamps, and in the village are two cedar mills. To the north is the Mullica River and Batsto. Practically every built and natural landscape of the Pines provides Landy with some kind of material. From the uplands come the acorn sprays from scrub oak, while finger grass and fern seed are found in pitch-pine lowlands. The high ground of the seashore provides bayberry, and the low cranberry bogs, the fluffy Indian and coffee grass. Old upland fields abound with wild wheat, dock, immortelle, foxtail, millet, thistle, brown burrs, mustard, and false heather. The salt marshes produce sea grass and statice, while fresh bogs and marshes have thousands of cattails, floral butts, fern leaves, and dusty gray reed. The fruit orchards of Hammonton supply orchard grass, and the ditches of the blueberry fields give fox plume, spirea, baby oats, and Joe Pye weed. Feather grass grows around old homesteads, cornettes in cornfields, and Douglas fir cones, gum balls, and white pinecones in domestic yards. The pepper grass prefers strawberry fields, and from the Plains come pitch pinecones.

PHOTO 18. *Leo Landy at His Pine Cone Kiln*   ©*James F. Gandy, Jr.*

   Access to these areas depends on good relations with landowners. For years Landy and his wife routinely asked permission, but after some years people just waved them on and wished them luck. In some cases the collecting is seen as a service that rids lawns of debris and orchards of unwanted grass. But some new landowners have no sympathy or understanding, even though they cannot use the materials themselves. In some instances Leo is barred from state lands; at Lebanon State Forest he cannot gather pinecones from the white pines he planted as a worker for the Civilian Conservation Corps in the 1930s.

Leo and Hazel provide seasonal employment for a number of people. She sends out catalogs to customers, and he makes contact with gatherers to supply them with materials. The Landys pay by the piece, and certain people specialize in particular materials. From Chatsworth come pitch-pinecone gatherers, and from Nesco and the tidal rivers come cattail cutters. Another man makes cedar boxes from sawmill wastes. At the start of the spring and summer collecting season, Hazel worries whether they will be able to fulfill their obligations to customers because they may have orders for as many as several thousand cattails. Leo never seems to worry because people somehow always show up to work. In the Pine Barrens there is a saying that "you can always make a buck on the bay and a dollar in the woods." Despite help, the Landys collect much material themselves, and they are completely responsible for the treatment and curing of all collections.

It is easy to cut the materials, but if they are not adequately cured, no florist will buy them. Modest outbuildings behind the Landy house belie the sophistication of the curing operation. A greenhouse made from polyethylene stretched over frames is a drying area for the full range of plants. Each plant has its optimum drying time, and the Landys tend each batch in an appropriate manner. Some batches are aligned in different places and on various angles inside the sunlit room to take full advantage of the heat supply, others are hung in bunches from the rafters, and still others rotated and protected from too much heat. At some distance from the "plastic house" sits a blackened building that is the pinecone oven.

Cones from serotinous (closed-cone) races of pitch pine open only with heat, and Leo receives them green and closed from the Plains. He roasts them to make them open and drop their seeds, and when open and brown, they are the attractive cones customers like. Pine seeds are collected and sold to seed dealers. Other curing processes provide magnolia, birch, maple, and azalea seeds to seed dealers. The last outbuilding is a packinghouse used for assembly of bunches and for storage. Here on open shelves that reach the ceiling are rows and rows of cardboard boxes filled with dried and bunched materials. Roughly scribbled cardboard placards line the walls, and a handmade fifty-five-gallon-drum stove heats the building. Bales of pinecones line the floor, cedar baskets crowd the corners, and various sprigs hang from the ceiling. There is a helter-skelter look, yet inside each box is exquisite organization. In small bunches are the ready materials beautifully cured and prepared, and a glance into each box exposes one to an array of textures, colors, and smells perhaps more impressive than the formal exhibitions at a flower show.

The Landys supply over two hundred florists throughout the eastern United States. These days they rarely deliver, but rely instead on parcel services or direct pickup by wholesalers. Large-lot orders for cattails, statice, and pinecones go to wholesalers in the Philadelphia region. To many of them, Leo is "just a local Jersey man who supples the cattails." Only the most abundant Pinelands plant materials find their way to large wholesalers, and at these big warehouses, the pinecones, acorn sprays, and cattails make up only a small percentage of the total stock. Yet this reliable market keeps the collectors employed. Smaller lots go to dried-flower retailers all along the East Coast, and a flower store in Boston may very well sell reed plumes collected in the Jersey marshes. For the florists who sell dried arrangements, autumn and Christmas are the major seasons, but the Landys have a year-round schedule of collecting and curing.

In winter some of the Landys' people trap, but there are those who cut birch whips in February that the Landys paint white. February is also the time to gather sweet-gum burrs off the lawns in Hammonton, and in March Leo travels to the sawmills to get scraps from the winter's cuttings, while Hazel sends out catalogs. April is the month for traveling the back roads to spot productive ditches and burns. Leo says, "I can feel a weed a half-mile away." In May and June it is time to pick pepper grass; July is for cattails, and August leads into the late summer and fall harvest season for almost all plant materials. After Christmas Landy and his wife can get some rest.

Leo likes his job because "I can go where I want to, when I want to without some boss telling me what to do." But the Landys are getting on in years, and sometime they will have to give up their business. Neither is sure if any of their children will take up the trade, but there seems little doubt that someone will continue the business because as long as the land remains open, there will be locals who will follow the seasons and collect the materials. Leo himself got the business from a man who had become old and lost interest, and if someone younger took over, the trade could be expanded to the live-plant business that the Landys abandoned a few years ago. Christmas trees can be grown, firewood cut, and decorations like pinecone trees, collages, grave blankets, and wreaths prepared.

Natural materials from the Barrens turn up in strange places. Christmas decorations in suburban shopping centers, floral arrangements in city law offices, Easter bouquets in downtown banks, and reforested hillsides along interstate highways are just a few outlets for the seasonal patterns of the woodland culture.

COLLECTING

The difference between gathering and collecting is that the former sustains one's life while the latter enriches one's data, specimens, and experiences. Insiders generally gather plants and fruits, but outsiders collect plants and animals. Over the past hundred years a small army of botanists, zoologists, and foresters have collected in the Pine Barrens, some of them with international reputations like Gifford Pinchot, Roger Conant, and George Woodwell. The reasons why so many scientists collect were described by the naturalist Carl Kauffeld (1957, 11):

> The urge to hunt and collect is strong in most of us, but never stronger in any one group of people than in naturalists—the zoologists and botanists. However much they may be immersed in fascinating details of a research problem, none hesitates to cover his microscope, close the specimen jars, place the study skins back on the shelf, or whatever—gather up collecting paraphernalia, and take off to any region that time permits. Sometimes this might be for six months or a year, sometimes only for a long weekend. All of us have the same enthusiasm for "field work"—the anticipation of seeing the plants and creatures in their natural state, whether this be only a few miles from home during our "day off," or thousands of miles away on the other side of the globe. Usually the more remote the better. The attraction of the strange and unknown is undeniable, but there are still many thrills to be had close at hand.

Kauffeld, an ardent lover of snakes, grew up in Philadelphia and got his first look at Pine Barrens reptiles when he met the most famous of Piney gatherers, the late Asa

Pittman, known as Rattlesnake Ace. It was through the insider's eyes that the outsider found what he wanted. As Kauffeld (1957; 36–37) described it:

> As early as 1928, when I was still a schoolboy in Philadelphia, Ace directed me to Mount Misery [rattlesnake] den. It was rather a trick to find in those days. There was no Route 70 then. The flag stop on the railroad at Upton, where Ace and his family were the sole inhabitants, was at the end of the gravel road that connected with the Brown's Mills road several miles to the north. From the Pittmans', all roads to the south, east and west were the double-rutted sand tracks that meander in the woods and swamps over the entire Barrens in an endless Labyrinth. . . . Mount Misery was one of the beauty spots of the Barrens, but the years have dealt cruelly with it. It is no longer undisturbed. In the quarter of a century or more that I have known it, it has changed from a remote spot far back in the forest from the traveled highways—haunted by ghosts of earlier times—to a mutilated area, first ravaged by the CCC and later by a summer colony of religious revivalists. A broad concrete highway passes virtually by its back door.

It is important that the reader understand the picture Kauffeld has drawn, for it applies to a small and important group of people who want the Pines to resemble as much as possible those remote areas of their childhood in which they took their most exciting trips—and, of course, the more remote the area, the better the chance that its specimens have not been collected.

Collectors in the Pines have produced both positive and negative results. Many scientists of major repute have contributed to efforts to preserve the Pine Barrens from major development, and as of this writing, at least forty-five research projects are under way on many aspects of Pinelands natural history (Bucholz and Good 1982). In so doing, they have publicized the Pines as a place in which unique plants and wildlife live, thus putting even more collecting pressure on rare and endangered species. Amateur collectors and those who collect for profit to sell to pet stores or the black market have devastated populations of rattlesnakes, corn snakes, and pine snakes. Clearly, the intent of scientists was not to further endanger rare species, but this threat is a natural result of publicity. This is a hazard common to preservation efforts; for example, the moment the federal government designates a wilderness area, hikers crowd into it if for no other purpose than to see what a wilderness looks like.

Last, there tends to be a conflict between the aims of scientists who want to preserve and residents who want to use resources. Some scientists have their egos invested in Pinelands preservation because their reputations were made on research in the region. Others have put much energy into saving the Pines on the national level but little into the slow, dull work of implementing management plans or fighting political battles at the local, county, or state level.

Only in rare instances has the understanding of insider and outsider been fused in a scientist's view. David Fairbrothers, professor of botany at Rutgers, is one of the best examples. Born in the 1930s in Absecon, a bay town near Atlantic City, Dave grew up not far from Don Zehner on the bays, and he was a clammer as a teenager. He produced several major publications on the flora of New Jersey and was one of the forces behind New Jersey's attempts to classify and protect threatened and endangered species, particularly in the Pines. But Dave does not see the region as a wilderness to be preserved intact.

"What we need," he said at a conference in 1978, "is directional change to preserve the Pine Barrens." What he meant was that people must plan disturbances to Pinelands ecosystems in order to set succession back to early stages, to preserve the patchwork patterns in the forest and create habitats for plants and wildlife. Cutting, burning, gathering, and hunting are as natural to the Pines as curly-grass fern and pine snakes.

Hunting

Surely the most popular activity in the forests, if not the most remunerative, is hunting. Some people hunt upland game birds, mainly bob-white and ruffed grouse, locally called partridge, but most hunters seek rabbits, fox, or deer. Most local hunters of upland game go for rabbit, mostly as a way of working with their beagles. For four months, from October through January, one can hear packs of beagles baying on the edges of fields and forests, and unofficial competitions among hunters and packs are common.

FOX HUNTING

The sole purpose of the fox hunt is to work with one's dogs. There are loosely organized fox-hunting groups everywhere in the region, and, while their numbers are not large, they cover almost every forested acre over the fall and winter hunting season. The fox hunters of the Pines are not like the ones wearing scarlet coats and white breeches in nineteenth-century prints.

> In the shadowy outline beyond them [the Redcoats], outnumbering them a hundred to one, are the legions of foxhunters, like Franciscan Brothers, whose profession of faith neither poverty nor sacrifice can dim, some who must even deny themselves the necessities in order to keep a couple of hounds. On horseback, on muleback, or more often afoot, every night of the year, somewhere in every state in the Union, the horns of this great army of hilltoppers awaken the echoes of field and forest. (Houghland, as quoted in Hufford 1982; 222)

As Donald Jones, an octogenarian hunter from Burlington County, put it: "They're the upper-class hunters, and we're the lower-class." Jones, like most of the hundred or so fox hunters in the Pines, does not kill or "take" the fox, but enters his hounds in competitive hunts: the hounds become an extension of their owners. Two or more hunters begin by setting their hounds on a trail they know foxes use. The hunters listen carefully to the hounds and can tell by the hounds' voices how close they are to the fox and whose hounds are ahead in the chase. Because the hunters know their territory intimately, they can jump in their pickup trucks and drive to the next point on the sand roads where they guess the hounds will cross. The idea is to beat the pack to that crossing in time to see the fox and to test which hunter was right about whose hound is first in the pack and what kind of fox they are after—red or gray (figs. 9, 10). The hunt stops when the fox tires and will clearly be caught or when the dogs trap the fox. At that point the hunters "break" the hounds from the fox's trail and go home.

FIGURE 9. *Gray Fox*

An extraordinary symbiotic relationship has developed over the years among hunters, dogs, and foxes and the patchwork pattern of the pine forests. If there are hunters and trappers who take fox for an income, they are more than matched by those hunters who see to it that the fox population is healthy and their diverse habitat needs met.

Of all the participants in the fox hunt, the most important and interesting are the dogs, whom the fox hunters refer to in anthropomorphic terms. As Mary Hufford (1982, 223–25) points out, the hounds' sole reason for being is to produce "music" for the hunters.

> The single most important feature on a hound is his voice and how he uses it. One rarely hears it referred to as "barking," but rather as "tongueing," "yelling," "squalling," and the like. The voice of the hound is the scoreboard of the hunt. It tells the hunter which stage of the hunt is in progress, and the position of the hound relative to other hounds. During the initial stage of the hunt, known as "cold-trailing," hounds emit long, drawn-out notes. When a fox has been started, they switch to a short chop. Hounds closest to the fox "chop" the hardest. Listening, the hunters may argue over whose hound is running "under the hammer" (that is, closest to the fox).
>
> The effort to achieve a full-bodied sound [from hounds] with room for individual definition, is reminiscent of an emphasis on self-sufficiency and democracy, which occurs repeatedly in conversations with these men. As working dogs, hounds are susceptible to unemployment. To earn their keep, they must perform up to a certain level of competency. Definitions of competency vary radically depending upon whether hunters catch the fox or protect it. For those who no longer kill foxes, a hound who will hunt to catch is considered to be too competent, and will not measure up to their esthetic standards. Two rules that may be violated both by incompetence and overcompetence would be as follows:
> 1) Hounds must make sense to the hunters.
> 2) They must follow the trail.

FIGURE 10. *Red Fox*

The incompetents are those who ''yell'' at anything (known as ''babblers'') and those who are not ''deerproofed'' (that is, they will veer from the trail to chase a deer or a rabbit). Those who are overcompetent offend by ''cutting'' (also called ''swinging'') and ''drying up,'' that is, they will ''wind'' a fox (catch his scent) and follow the scent, which is usually in a more direct line with the fox than the trail. To keep the hunters from knowing, and to gain an advantage over the other hounds, they will not signal. They become so involved in the competition that they usurp it and become unaccountable to the hunters.

And what is the point of this ritual? Hufford asked the question of Donald Jones. ''Listen,'' said Jones, ''they're talkin' to you!'' ''What are they saying?'' asked Hufford. ''They're tellin' you the fox is ahead of 'em.''

DEER HUNTING

By far the most popular hunting in the Pines is for white-tailed deer, and not only insiders participate, as in foxhunting, but outsiders as well. As with fox hunting, the patchwork pattern of the Pine Barrens is a fitting landscape for the activity, and most men born in the Pines grow up knowing how to hunt.

Bill Wills, Jr., is now almost thirty. His dad, Bill, Sr., is almost seventy, and the Wills family has lived and hunted in the Pine Barrens since the late eighteenth century. Young Bill's first memories of hunting were of his father bringing home game—rabbits, clapper rail, grouse, deer, and waterfowl. By the time Young Bill was twelve, the age when most boys go out on their first deer hunt, Old Bill was in his fifties, and like many men that age, more interested in his job than in hunting. But Old Bill still had enough time and energy to take his son out deer hunting twice that year.

Of course, Young Bill knew how to handle a gun. His father had worked with him on target practice with a .22 since the age of nine or ten. Age twelve was the first time he took out the twelve-gauge shotgun and shells filled with buckshot and went out with his dad to shoot a white-tailed deer. Since Young Bill had grown up in the Pines, he already knew a little bit about the habits and patterns of the deer and, with the help of his father, where deer could be found that day.

In a ritual that Piney men have enacted annually, Young Bill and Old Bill worked the woods together. They stepped quietly through the forest in which they had both grown up, stopping occasionally where deer trails crossed or where they had recently seen a deer or its sign—antler scrapes on a tree, new tracks, droppings, and clipped browse. They listened intently for several minutes and moved quietly along the edges of the cedar swamp and blueberry field. Their route was well chosen because they walked the edge of the deer's preferred winter habitat.

White-tails are most numerous in areas where such edges are available and where mixed vegetation occurs. Disturbed sites are most productive, because the open canopy lets in sunlight, which encourages shrub and grass production, staple items in a deer's diet. They like succulent sprouts in newly burned and open areas like blue flag, turkey beard, sedges, reeds, and grasses, especially in spring and summer when deer graze rather than browse. In fall, organisms in a deer's rumen and intestines change, so deer browse during late fall and winter, and white cedar is the most preferred vegetation. But the difference between a healthy and a sick deer herd in the Pine Barrens is usually the availability of mast, especially acorns from scrub oaks.

Acorns are highly concentrated food; they are necessary because quality is more important than quantity in the protein-deficient diet of the Barren's deer herd. When deer cannot get enough acorns, they feed on crops and less nutritious browse. In low-nutrient diets the quantity of food needed to fuel the animals' energy requirements is larger than the digestive system can handle. When deer exist on a diet low in quality and digestability, their energy intake is below their maintenance level, and they get inadequate supplies of calcium, phosphorus, fat, and especially crude protein. During a poor season, deer use up all their stored fat, fawns may die because they were too small to store the necessary reserves, bucks may not grow antlers, and does may resorb their fetuses that were conceived in the fall. Without emergency acorns, deer often starve with stomachs full of pine needles and oak leaves.

The scrub or bear oak (*Quercus ilicifolia*) provides most of the Pine Barrens' acorns (Wolgast 1977, 1978). A period of low humidity at the time of flowering generally insures a good autumn crop, yet oak trees on the same site may have different volumes of acorn production; genetics may even play a role in production. A smart hunter can pick good deer spots if he knows the locations of good acorn crops.

To the Willses and to the deer, the Pine Barrens are not an unbroken forest, but a recurring patchwork of small quadrangles bounded by natural and man-made edges. The one- or two-foot differences in topography and, thus, water table produce alternate wet-dry patterns, and each transition is an edge that provides food and cover for deer and landmarks for hunters. The cedar swamp has its white cedar, the border hardwood swamp or thicket can provide sassafras and even scrub oak in its drier sections, and open pine woods can grow food for grazing. When fire burns these areas and arrests succession, more edges and more food sources occur.

When the Willses go into the woods, they know that along every edge is a main deer "avenue" with many small branches. The meanders of riparian vegetation provide a series of sinuous trails, and the rectangles of blueberry fields are templates for a grid system. Trails may lead from cover to food; a small rise covered in laurel and oak and incised by small hollows may hold little food but provide shelter from the weather and bedding for sleep. Downslope from the rise is a recent burn, dotted by pitch pines and containing browse and turkey beard, and at the edge of the hollow is a small spung that runs parallel to the burn. This is a perfect situation for the hunter and deer because trails come off the rise, one from each small laurel-covered hollow, and converge on two major avenues heading for the turkey beard. Arteries branch from the main routes and head toward the cedar swamp, the best protection from hunters.

On that day when Bill went out with his father, they flushed deer from hollows on the rise. As two small bucks jumped out of the laurel, Young Bill shot his first tree limb with his twelve-gauge. It was as it should have been. Any hunter knows that on the first day it is easier to shoot a tree than anything else, and every Piney man can clearly remember the day he shot his first branch.

After Young Bill's first hunt with his father, he frequently went hunting alone for small game. He came up empty-handed his first year and, perhaps because of these disappointments, did not hunt when he was thirteen. The year off merely whetted Bill's appetite, and he went hunting with enthusiasm when he was fourteen. He has not missed a year since then. Most of the time Bill, Jr., hunted alone, jump-shooting small game by carefully stalking good cover. His first kill that year was a Canada goose, of which he was quite proud. Since small game and waterfowl are much more abundant than deer, the small-game season lasts longer than the deer season, and Bill spends 80 percent of his time hunting small game. But in the Pine Barrens hunting prowess and excitement are measured by the remaining 20 percent of the season which is Deer Week.

Hunting is as much a part of life for Bill Wills as it is for most men who live in the Pines. Unlike outsiders, for whom hunting is a break in their normal pattern of work and home life, Pineys integrate the hunting season with their annual life-cycle. Instead of a hiatus, hunting is a continuation and reaffirmation of an annual pattern with which people like Bill grew up. It is not a pedestrian workaday chore like a job. It constitutes a special part of the annual cycle, which, like marriage or work, is more important to some men than others and around which stories, rituals, and songs have grown.

Except under extreme conditions, like the Depression, deer are not a crucial part of the diet of Pineys. They gather directly from their gardens and earn cash, so game is a dietary supplement, not a necessity. Even under optimal conditions, the relative abundance of game could never support more than one person per square mile, even if that person killed six deer and a significant number of rabbits, waterfowl, and other game. Pineys also raised their own pigs and chickens until World War II; pork and garden vegetables carried many a family through the winter, and no old-timer can remember when gardens did not produce a good crop of something. The following song, ''Home in the Pines,'' by the late Uncle Bill Britten, expresses these sentiments. He wrote it in the early 1940s, and his niece Janice Sherwood and the Pineconers still sing it at the weekly Saturday Night Jamboree in Waretown (Gillespie 1980, 35–36).

When you settle in the pinewoods down by the Jersey shore
You're gonna have the hard time knockin' at your door.
When you plant yourself a garden against the time of need
You're gonna be lucky if you ever raise your seed.

So folks I'm gonna tell you if you crave to settle here
You needn't bring your city clothes along,
For whate're you turn your hand to make your daily bread
It's ten to one it's gonna turn out wrong.

The soil is doggone sandy, the insects try your soul
For first they eat your beans up and then they eat your pole.
The deer will leap your garden fence in the middle of the night
If you want to save your cornfield, you got to get up and fight.

So it's work all day and gun all night, there is no time to sleep,
Except when you're leaning 'gainst the rail.
And if you kill the pesky deer that's eating your beans,
The warden wants to throw you in his jail.

Now when you go and dig some clams to sell when you reach shore,
You'll find that the people just don't eat clams no more.
But when the season changes and clams just can't be found,
You'll find a buyer waiting each time you turn around.

Oh the deer eat up your garden and the clams you cannot sell,
Your luck just runs a little worse each day.
But you've got to stay and take it till the bitter end,
For you haven't got the means to get away.

But folks in spite of all this, there is something about these woods
That you will feel if here you ever roam
And when you see the sunsets and breathe the balmy air,
You'll never want to leave your Jersey home.

If the major purpose of hunting had been for food, game would have disappeared long ago. Hunting is, in fact, a social event, and most Pineys use hunting to educate children and foster father-son relationships. It teaches self-reliance, independence, and an intimate knowledge of the environment, all skills that perpetuate woodland society. When small groups of hunters participate in small-game hunts, they scout for deer. The month of November is still a traditional time to watch for deer signs and plan for Deer Week in December, the one social event that binds together men in the Pine Barrens.

If Pineys wait for Deer Week, so do the other thousands of hunters who have permits to shoot deer. The majority of these outsiders belong to hunting clubs (photo 19), and, unlike most Pineys, hunt in large groups called "drives." The influx of outside hunters swells the seasonal population of the woods to the point where many residents complain they do not dare step out the door for fear of being shot the first week in December. However, the gun clubbers, through their numbers and organization, have strongly influenced the very life of the white-tailed deer and the entire big-game hunting system.

PHOTO 19. *Pinewood Antler Gun Club*   ©*James F. Gandy, Jr.*

Since the seventeenth century, New Jersey has enacted game laws to protect wildlife resources through regulation of hunting. Despite such regulations to protect the diverse and suitable habitat for deer in the eighteenth and nineteenth centuries, hunters, especially those with dogs, succeeded in extirpating white-tails from most of the state by 1900 (Coggins 1980; Tillet 1961). In 1901 the total harvest for New Jersey was twenty deer, and the legislature closed the entire state to deer hunting from 1904 to 1913. During that time, deer from other states were introduced, and the deer population rebounded with a vengeance. With all predators eliminated, neither man nor automobile could check population growth, and since the 1930s deer herds have remained at or close to "carrying capacity" (the ability of a region to carry a given number of animals with sufficient food, water, cover, and living space for them).

In 1945 the state legislature created the Fish and Game Council, which, to this day, has the authority to establish and administer with scientific advice a game code regulating seasons, types of weapons and ammunition, hunting hours, bag limits, sex of animals taken, and even clothing requirements. The governor appoints to the council: two sportsmen and one farmer from each of three districts (north, central, and south), one Department of Environmental Protection employee who works with nongame and endangered species, and one person from the U.S. Soil Conservation Service skilled in land-management techniques.

Members of hunting organizations are a majority on the Fish and Game Council and, thereby, exert a strong influence on wildlife management and the use and preservation of open space. Their influence stems from the revival of the deer herd, which prompted a rise in the population of hunters who founded a statewide sportsmen's federation in the 1930s that sends members to the council. Even though sportsmen dominate council voting, they do not speak with a single voice because arguments concerning hunting laws are complex, and bitter feuds break out among hunters about proper regulations.

Immediately after World War II untutored wildlife managers and some parts of the scientific community believed that new food sources, planted by wildlife managers, would divert the attention of deer, which were destroying food crops. They saw no need to change the age and sex composition of the herds. Supporters of the ecological views of Aldo Leopold contended that any given range had only enough food, water, shelter, and living space for a given number of deer regardless of new food sources. One could choose, therefore, between a large number of unhealthy deer that would stress their environment, or a smaller, well-fed herd that would produce more and healthier offspring. The population-control group recommended a doe hunt to reduce the size of the herd and maintain optimum carrying capacity. Hunters were split on the question. Some believed the food sources were adequate to protect property and maintain the herd, but a larger number believed that there were plenty of big bucks and it was natural for some deer to die of starvation (Tillet 1961). Further, anyone who tampered with the game laws violated the rights of hunters who paid for the entire wildlife-management system through a dedicated fund derived from license fees and taxes on equipment sales. The hunting of does, they said, would not solve a nonexistent problem, and, besides, hunting doe was cowardly; to this day many hunters will not kill a female.

In the Pine Barrens many gun clubs supported the "bucks-only" policy. Crop losses in the Pines were small compared to those in North Jersey, so the deer-damage argument was less compelling. Moreover, the range capacity of the Barrens may have deceived the hunters. Through the thirties and forties hunters benefited from the last remains of edges created by large forest fires, but mechanized fire suppression began to work after that, and late successional stages began to eliminate the open clearings and their many food sources. The hunters may not have realized the changes in the forest and could not perceive that the number of deer would decline as habitats deteriorated. They were used to high populations of deer, which they attributed to the prohibition against hunting doe rather than a beneficial environment. The "save-the-does" policy also sat well with nonresident hunters whose lives were not generally affected by deer populations. Southern representatives on the Fish and Game Council held firm against a doe season, while northern members, beset with a variety of complaints, were the first to vote for doe hunts in their counties. In the late fifties and early sixties the controversy continued to rage.

At the height of the controversy, professional wildlife managers for the state, who advised the Fish and Game Council, found themselves in a dilemma. A growing body of evidence suggested that a combination of habitat manipulation and bag and sex regulations was the best form of management; yet many hunters disagreed. At the same time, demand for the state's wildlife resources seemed to increase in direct relation to the decrease in habitat caused by urban growth. Managers were caught between the intransigence of their chief constituents and the need to develop more effective management strategies.

To add to their predicament, the managers by law could influence only population-control policy, not land-use patterns. The game code was their only effective management tool, and they had few powers to manage the habitat of deer. Had they wanted to implement a different fire-management, forestry, or zoning policy, they could not have done so directly. Had they wanted to curtail the loss of farmland and woodland, they would have had no financial resources or political power to do so. In their view it seemed that most recreational hunting would eventually occur only on public lands because of the loss of private range and the increasing number of hunters. Against popular opinion and without the evidence they needed for an airtight case, wildlife biologists continued to press for doe hunting to control deer herds in the face of alarming decreases in the weights of fawns and decreasing antler size on bucks.

The hunt clubs, through the Sportsmen's Federation, continued to argue with the biologists; by law the Fish and Game Council had to act on scientific advice, but the scientists lacked data. It was only in the 1970s that wildlife biologists began to gather information precise enough to estimate deer populations and habitat condition so they could present a more solid case. The establishment of mandatory deer-check stations and the experimental North Jersey doe season led in 1974 to the adoption of a statewide either-sex deer season based on small-deer management zones.

P. K. Hilliard of Pleasantville, near Atlantic City, remembers the bitter fights about deer management during his ten-year term as a member of the council in the seventies. His South Jersey constituency was solidly against the suggestions and ideas of the wildlife biologists regarding such issues as carrying capacity and doe seasons. At first Hilliard agreed with the gun clubbers, but then he started to read the results of wildlife surveys carefully and became interested in the controversies. In the end, he supported the views of biologists and told the sportsmen, "If they had a better way [than the biologists'], they should submit it; if not they should support the proposed system." The hunters finally voted for the biologists' plan, and, as Hilliard recalls, "the old guys died hard. But when the Bureau [of wildlife] put out the prediction of harvest based on the number of permits, and when the prediction came through, many people came over."

Since then, hunters and the council have cooperated, and the system appears to work quite well. In 1980, for example, four either-sex seasons were added to the traditional Deer Week. Due to the low productivity of Pinelands forests, deer zones in the Pine Barrens carry fewer deer per square mile than other sections of the state, and the permit quotas to hunt in the region reflect that difference. Zones 23 and 26, north and south of the Mullica River and favored by the gun clubs, have a minimum average population of 9 to 12 deer per square mile in predominantly pine and oak forests. In 1979 hunters harvested 335 bucks from zone 23 and 340 from zone 26. In contrast, zone 21, dominated by the Plains, has an estimated 6 to 9 deer per square mile, and in 1979 hunters killed 293 bucks. Hunters who expect to succeed must know their region well.

Hunting success in the patchwork mosaic of the Pine Barrens depends heavily on scouting before the hunt. Danny Franchetti is a captain of the Never There Gun Club of Hammonton, and he contends that "good preparation is important, but you never know what's going to happen when the deer moves. The deer doesn't know, and the hunter doesn't know.''

For Danny, Deer Week begins three or four weeks before the opening of the gunning season, during late fall, when he walks and rides the back roads to check for deer. He looks for all signs of deer, checks to see which lands are posted, which old and new roads are open, and whether there are new houses in the club's traditional hunting territories. He wants to find out the schedules of the larger bucks—where they bed down and where and when they feed. Like the Willses, Danny scouts all the edges and patches. He may increase access to areas by blazing lines laid out by compass through the forest. Deer drives must be planned, precisely because it is easy for a drive to disintegrate if hunters lose their orientation and begin to walk in circles.

Twenty-seven men, equally divided between drivers and standers, usually participate in a Never There drive. Drivers are usually a half-mile from standers, positioned by Danny according to the latest scouting and wind direction. The drivers' responsibility is to walk straight ahead, in a close line, always in visual contact with each other. Standers are placed upwind, often on roadsides, and deer, when flushed, often quarter upwind. No matter how bad the weather, the standers are not supposed to move from their stations. The buck's strategy, too, is to crouch in hiding, not moving if possible. Even if the captains know there are two or three large bucks that use an area, the hunters may not be able to see the deer as it quietly walks behind a laurel bush or backtracks away from them. If the drivers can flush a deer, they should get a shot, and most bucks, in fact, are killed by drivers. But if they miss or cannot get a shot, they hope the deer will run toward the standers who, if they see the deer, can sometimes get a decent shot, if they are not too cold, wet, or bored. The standers may have been waiting a good while before they hear shots. If they hear the drivers, they know they cannot shoot in front, nor can they shoot to the side because of the other standers. They could be caught completely offguard when a buck explodes across a road, or they may miss seeing the deer as it quietly slips by. After each drive, standers and drivers usually switch positions.

Despite the best scouting and experience, chance plays a large part. Although deer will quarter upwind, they will also try to escape in a swamp. When the wind and swamp are on the same line, the standers can guard access to the swamp; otherwise, Danny has to choose which area to cover or select another spot. When he scouts, Danny looks for different-sized ''blocks,'' or hunting sections. He may choose the block because of wind direction, scouting information, or simply the number of people on that particular drive.

A whole series of rituals and regulations attend gun-club activities. Safety and sporting conduct are inviolable. Members cannot ''bait'' an area; that is, they cannot leave apples or corn within three hundred feet of a tree if they want to wait in a tree stand for deer to pass underneath. They must make every possible attempt to find a wounded deer, and all deer must be checked at the state checking station. Each club also has its own social rules, its favorite foods, and its special nights. Every club has its own history, and its special stories are told over and over.

Deer Week is a major time commitment, and the activities are well planned. No club is run democratically: older members and captains organize the drives, the division of labor, and the budget. Hunting hours are set and adhered to. Many clubs hire cooks, and Never There members have a substantial lunch brought to them in the field at noon. Dinners can be enormous and the social life memorable. At the end of Deer Week, on Saturday, the kill is taken down from hooks in front of the clubhouse, wrapped, and stored for the winter game dinner attended by members and their families. The men return home on Sunday.

Most hunters leave the Pines and return only at the start of the next deer season, but some use the club throughout the year. Sometimes club privileges are extended to area residents. Never There hunters come from Hammonton and are the exception to the rule that most gun-club members come from outside the Pinelands region. Their clubhouse is not open in summer.

When Deer Week ends, Danny Franchetti no longer searches the edges and hollows for deer. All that remains is to pester the club members for their dues. He will leave the sand roads, the swamps, and the uplands to other users. His scouting is done until next autumn. However, he and his fellows look forward with anticipation to the next season and feel their deer harvest will help keep the herd healthy. There is pride in their accomplishment. One hunter commented at the end of a hunt, "We take something out of the Pines, but we leave something and we give something back." He added, "Those people who know the woods always give something back."

Insiders and Outsiders, Part 2

We have now come to a point where we can more fully appreciate the battles between insiders and outsiders and their impacts on the landscape. While Danny Franchetti lives for the most part outside the forests of the Pine Barrens, he treats the forest as an insider would, by giving something back when he takes something out. The insider, then, is not just someone who was born in the forest and understands the politics of the place, but a person who gives of him or herself to that place and is on intimate terms with it. Aldo Leopold (1949, 203–10) put it this way:

> All ethics so far evolved rest upon a single premise: that the individual is a member of a community of interdependent parts. His instincts prompt him to compete for his place in that community, but his ethics prompt him also to co-operate (perhaps in order that there may be a place to compete for). . . . In human history, we have learned (I hope) that the conquerer role is eventually self-defeating. Why? Because it is implicit in such a role that the conqueror knows, *ex cathedra,* just what makes the community clock tick, and just what and who is valuable, and what and who is worthless, in community life. It always turns out that he knows neither, and this is why his conquests eventually defeat themselves. . . . No important change in ethics was ever accomplished without an internal change in our intellectual emphasis, loyalties, affections and convictions.

We have seen a number of outsiders in this narrative who have become insiders, among them the De Cous of the lake community, the Wynns of Egg Harbor, and Jack Cervetto of Warren Grove. They accomplished their transformations by giving back to the place as much as they took. The gun-club members, for example, give to preserve the patchwork pattern of the forests. Of those who give their livelihoods to their place, no better example exists than the Albert brothers, who began the Saturday-night jamborees and were founding members of the Pinelands Cultural Society.

Joe and George Albert were born in the second decade of this century in Sayreville, Middlesex County, one of the many sphincters on the rivers of northern New Jersey. The brothers began hunting deer in the Pine Barrens of Ocean County in the 1920s and built a hunting cabin in 1933 in the pinewoods ten miles west of the bay community of Waretown. Joe moved to the cabin in 1939 after a string of disheartening jobs on the Raritan River, and George and his family regularly visited Joe on weekends, when they would play music together. Other musicians joined them, and, when George retired in the late fifties, music making became a fairly regular Saturday-night occurrence that turned into the Saturday-night jamborees at what became known as "The Old Home Place," the Albert brothers' cabin.

In the mid-seventies the Old Home Place became so popular that teenagers gathered in droves on Saturdays, and their rowdy drinking became a problem. The jamborees had to be moved to a more conventional setting. George died in 1973, and the musicians, their friends, and their families established the Pinelands Cultural Society in 1975. They began to give the jamborees at the auction house in Waretown, which they dubbed "Temporary Albert Hall," fully aware of the ironic implications of the name. Joe still hunts fox, but instead of hunting deer, he feeds them, and one can see a score or more deer at the cabin on any winter evening. And he still plays washtub bass and sings occasionally with the Pineconers at jamborees. The Pinelands Cultural Society in 1981 had gathered sufficient funds from dues, concerts, and food sales to break ground on a piece of land given them. Permanent Albert Hall is under construction (Ayres 1979).

The activities of the Pinelands Cultural Society are still concentrated in central Ocean County, a section of coast and Pinelands that is hard pressed by suburbanization. The society concentrates on folk music as a rhetorical strategy to preserve the local land ethic and to attempt to make insiders of the outsiders who would suburbanize the landscape without regard to the ethics of the place. As Janice Sherwood says (Ayres 1979, 229–31):

> We're just a little bit cynical and a little disappointed at the way some of the things have gone, and if we could put up our hand and hold back time, we would be very glad to do it. But you have to be realistic, you know, that it isn't going to happen. We would like to maintain some of the standards and attributes that we grew up with. And we feel that music is one of the ways to do it—to lead people to thinkin' a little bit our way, and not necessarily about—well, of course, ecology has a whole lot to do with it. . . . We do not like to see our streams polluted any more than they have to be. We do not like to see our woods carpeted with garbage. We do not like to see gravel dug out in great gouges where pines used to be. . . . There's so many people—well, maybe they wouldn't want to destroy so darn bad if they had any concept of—if it meant anything to 'em. A pine tree doesn't mean anything to 'em, or a holly bush, or the mountain laurel when it's blooming in the spring. It doesn't mean anything until these people, until they really get a chance to hear about it an' see it an' feel it.

Notice the realism tinged with resignation in Janice's words, "We do not like to see our streams polluted any more than they have to be." She understands what a balanced, as opposed to a perfect, landscape means.

If one does not need to be born in the Pines to be an insider, so one can come from an old Piney family and become an outsider. We have seen that conflict especially in suburbanizing areas. Likewise, in the forests and fields live people who leave, never to return from the cities and others who speculate on land. Some farmers and berry growers would sell their heritage at the right price, although most would not.

The long-term health of the Pine Barrens and its people will be decided over the next five years. Aldo Leopold's advice still holds (1949, 214–15):

> A system of conservation based solely on economic self-interest is hopelessly lopsided. It tends to ignore, and thus eventually to eliminate, many elements in the land community that lack commercial value, but that are (as far as we know) essential to its healthy functioning. . . . The "key-log" which must be moved to release the evolutionary process for an ethic is simply this: quit thinking about decent land-use as solely an economic problem. Examine each question in terms of what is ethically and esthetically right, as well as what is economically expedient. A thing is right when it tends to preserve the integrity, stability and beauty of the biotic community. It is wrong when it tends otherwise.

The words of the 1979 New Jersey Pinelands Protection Act speak of the right things—of the integrity and preservation of the biotic community and indigenous industries. Can water be kept as clean as possible, not necessarily "pristine"? Can farmers continue to farm? Can the patchwork pattern of the forest continue? The question for the last chapter remains: Can a plan be written that reflects that balance inherent in the voices of the consciences of the communities?

Chapter 5

RESPONSE

We have looked at the Pine Barrens both from within and without and discussed point and counterpoint, topophilia, and a host of activities related to resource use. We have emphasized views from the inside because these realities, most often neglected by planners, can give us a deeper understanding of the region, expand our planning alternatives, and lead to more effective planning. An understanding of socionatural processes can contribute significantly to a regional plan. We again stress that we are developing an aid to existing planning practices, not a whole new framework.

We first present a short history of planning efforts in the Pines, then a description of the planning process and problems of the Pinelands Commission, and finally a set of propositions that can help expand planning opportunities not only in the Pine Barrens, but in rural areas throughout America.

A History of Planning in the Pine Barrens

The planning process can be highly formal, as it is with township planning boards, or it can be informal, as with the decisions of individuals to use their property in different ways. Merwyn Samuels (1979, 74–76) calls this latter process "the authored landscape," and it is as common in the Pine Barrens as it is throughout the Northeast.

Landscape expressions, no less than impressions, belong to someone in some context. The fact of social plurality does not mean that shared landscapes have no authors. Neither does it mean that no one in particular can be held accountable for that landscape. . . . The makers of landscape imagery in the modern context are often the makers of landscape designation—real estate agents, brokers and developers are often good sources for a biography of landscape designation. Just as they convey and perpetuate landscape intentions cast by others, so too do they create, manipulate and designate the forms and meanings of places.

Much of what one sees on the landscapes of the Pine Barrens is the work of various individual authors, both known and unknown. Prior to World War II almost all land-use planning in the Pines was of the informal, authored type; no Pinelands town had a planning board, nor did the state have a planning agency, such as the Division of Planning in the Department of Community Affairs that presently exists. With few exceptions, the authors of the landscape were born in or had long experience with the Pines and worked with its resources, so land-use planning stemmed from organic connections between people and the land.

After World War II, land uses in the Pines, indeed throughout much of America, became both more specialized and more indelible. On top of the generalized rural landscapes were placed single-purpose developments such as suburbs, extensive gravel and sand quarries, and commercial-strip developments. These kinds of developments, besides excluding other land uses, had almost irreversible impacts. One cannot, for example, easily turn a suburb into a field or a shopping mall into a woods. With the advent of specialized landscapes and the absence of standards for proper development, the Pine Barrens began to experience irreversible destruction of its land and resources as residential developments sprang up in inappropriate places, such as fire-prone forests and swampy lowlands.

Townships were then suddenly left with expensive maintenance and service problems because roads and highways, even whole subdivisions, flooded annually, and unexpected demands for increased school enrollments forced tax rates higher. Long-time residents with few expectations for social services were pitted against newcomers who expected sophisticated infrastructures; old people refused to vote for new, sometimes necessary, schools for newly arrived young families. As tax rates rose, old rural residents often had to sell more of their land, which increased land speculation and development pressure. Most important, the extensive, contiguous open space, one critical feature of the Pine Barrens, could no longer be taken for granted. The informal planning that had resulted from old authored landscapes had broken down. Despite long-time residents' deep distrust of any authority, even on the township level, they accepted formal planning as necessary, although many towns held off until the late sixties, after parts of their land had been destroyed. Many, if not most, Pinelands residents still resent formal planning, and their anger is fueled by cynical outside developers eager to slap cheap, profitable developments onto any available vacant land. The dilemma is real: many residents want to retain the old landscape while at the same time being able to sell their land for the highest possible price to a developer. The mixture of conservative rural values, suspicion of outside authority, and love of the land with its attendant nostalgia creates a difficult climate in which planning must proceed.

Since 1960, people have attempted to plan for land use in the Pine Barrens on local, state, and federal levels. As is common in northeastern states, most planning and police, or zoning, powers reside with township governments. Until the last decade, the towns of the Pinelands were solely responsible for land-use planning. This institutionalized process is called "home rule," and planners cannot underestimate the importance of this tradition, which extends from colonial times. Local people may be suspicious of local politicians, but they say, "at least we know where to find them."

Unlike many places in the Northeast, however, the Pine Barrens have resources of regional and national importance, namely, the thousands of acres of open space near the megalopolitan corridor, the trillions of gallons of potable water underground, and the unique flora and fauna that are internationally renowned. These resources know no town boundaries, nor are they properties of individual townships; they need planning on a regional level.

Regional-planning efforts came on two fronts—the central Pine Barrens and the coastal zone. The first attempt came from the now defunct Pinelands Regional Planning Board appointed by the governor of 1964. Their 1965 report included four alternative plans, among them the infamous proposal for a huge regional jetport and town for 250,000 people in the heart of the northern section. From the outraged reaction, which killed the planning board itself, came the first major study that advocated a preservation plan.

Meanwhile national and state interest in coastal areas quickened. New Jersey enacted its 1970 Wetlands Act shortly after passage of the national Coastal Zone Management Act; the state act prohibited any disturbance of coastal marshes without a permit (those areas with cordgrass, or *Spartina spp.*). Its sister act, the Coastal Area Facilities Review Act (CAFRA), passed the state legislature in 1973 and granted the Division of Coastal Resources (formerly the Office of Coastal Zone Management) of the Department of Environmental Protection authority to regulate development of twenty-five or more units and major public and industrial facilities in the coastal zone. In New Jersey the coastal zone includes all coastal areas of the Pine Barrens and adjacent uplands. The Wetlands Act has successfully preserved the salt marshes, but planning for the coastal areas has had its friends and foes; some feel it restricts development too stringently, others that it allows too much. CAFRA has certainly proven more beneficial than detrimental in terms of resource preservation, and, while local towns and counties often grouse about the DCR's bureaucracy, they generally accept it as a necessary evil or even sometimes as a planning aid.

A year before passage of CAFRA, the state legislature created another regional planning body for the central Pine Barrens called the Pinelands Environmental Council, which was funded half by the state and half by the three counties of Monmouth, Burlington, and Ocean. The PEC had only review powers over major developments; that is, it could review developments and delay them sixty days, but could not prohibit destructive ones. In 1975, the PEC produced its Plan for the Pinelands, which Governor Brendan Byrne called a "developer's dream."

The years 1975–76 proved to be a turning point in the fight to prevent further suburbanization of the Pines. During those two years a number of threads were interwoven. First, local residents in the northern sections of the Pines became alarmed at residential developments that were ruining the character of their towns and straining capacities to provide services. Many knew their towns needed outside help to curb the destruction. Second, environmentalists in South Jersey coalesced to form a political force for Pinelands preservation. Third, federal and state officials maintained active interest in Pinelands preservation, particularly Congressman Florio, Senators Case and Williams, and officials of the Department of the Interior, especially the National Park Service and the Heritage, Conservation and Recreation Service. Fourth, New Jersey politicians, led by Governor Byrne, took an extraordinarily active interest in the Pine Barrens.

At Brendan Byrne's Pinelands conference at Princeton, in late 1976, he gathered insiders and outsiders to discuss strategies for preserving the region. The following May he established the Governor's Pinelands Review Committee (PRC) to collect data and suggest methods to implement planning and management strategies. The PRC lasted almost two years and was instrumental in preparing the way for New Jersey's Pinelands Protection Act; the organization and goals of the Pinelands Commission were based on PRC recommendations.

The following year, 1977, saw intense lobbying in Washington to produce Pinelands legislation. Environmental groups wanted legislation that would provide a powerful centralized regional commission to prevent development, while mayors and other local politicians sought a decentralized structure. The environmentalists won, and in July 1978 Congress passed the National Parks and Recreation Act, which established the concept of a national reserve. Section 503 of that act created the Pinelands as the first reserve for which the state would create a regional planning commission and match federal funds that were appropriated for planning and acquisition; the federal government was also required to supply expertise from the Department of the Interior whenever possible. The president signed the act on November 10, 1978.

The stage was then set for passage of parallel legislation by the state of New Jersey, and in February 1979 the Pinelands Protection Act was introduced into the state legislature. That same month the PRC produced its report, and Brendan Byrne signed his Executive Order #71, by which he appointed seven Pinelands commissioners and made all development in the Pinelands subject to review by the commissioners. The seven affected counties and the Department of the Interior also named their commissioners in February, and the fifteen-member Pinelands Commission held its first meeting in March.

THE PINELANDS PLANNING PROCESS

What happened over the next four years tells the story not only of planning for the New Jersey Pine Barrens, but of many regional planning efforts in America. (A full-length history of the Pinelands Commission [Robichaud-Collins and Russell, n.d.] is in preparation.) The Pinelands planning process, by no means evil in its origins or intent, nonetheless highlighted crucial problems we continue to encounter throughout the country.

To understand the planning process, readers need to keep in mind our discussion in the first chapter about the tasks and purposes of planners. The first task is to understand a place and its people, or, as MacKaye (1928) put it, to obtain "an accurate formulation of our own desires—the specific knowledge of what it is we want; and an accurate revelation of the limits, and opportunities, imposed and bequeathed to us by nature."

The second is to establish where conflicts and congruences occur, and the third, to use techniques that will help create as healthy an environment as possible. The first, or revelatory, task is important because it reveals the interactions between people and place. A planner's ability to understand the substance of a place is crucial to a recognition of what it is possible to achieve in the second and third steps. It helps minimize the inevitable conflicts between a planning/regulatory agency and the local population, many of whom naturally resent the intrusion of a government agency into their local affairs. It is the first task to which we hope our work contributes.

If one were to look at a flow chart or schedule of phases for a regional plan, one would generally find the following divisions: (1) The establishment and interpretation of the goals of the plan, which, in the case of the Pinelands, were accomplished by the 1978 and 1979 federal and state acts. (2) An inventory and mapping of the region's most important ecological, economic, and sociocultural features. (3) The analysis and synthesis of inventories, including cartographic interpretations of the most important features and an assessment of the most critical areas for preservation as well as those areas with the greatest potential for development. (4) The creation of the plan itself, which should be the product of the first three phases plus the incorporation of input from the public, who should have been involved throughout the planning process. (5) Implementation of the plan, which uses the tools developed in the fourth phase. The first two phases cover the task of revelation; the third uncovers conflicts and congruences, limits and potentials; and the last two phases help create what should be a healthy environment.

One can easily understand how interpretation of material during the first phase can lead to appropriate or inappropriate results later on. One must also understand that rarely is a plan carried out according to textbooks. If any one law governs planning, it is Murphy's Law, which states, If anything can go wrong during a planning process, it will. In fact some even more cynical planners have suggested that Murphy was an optimist! Working with conflicting goals, interest groups, and staff members, with enormous amounts of data, and under considerable time and political pressure is by definition a frustrating job. For it to succeed at all, planners must approach the job with a combination of an open, inquisitive mind and a sense of purpose, which is to help create as healthy an environment as possible in partnership with the people who use the land and its resources, both insiders and outsiders. Like a parent, a planner knows that some things are bound to go wrong during the process, but if the actions and intent of planners during the process are ethical, if the planners continue to learn from the mistakes that will occur, then the progeny will likely be healthy and will be able to mature appropriately.

The first six months of 1979 set the course of Pinelands planning, because the commissioners and staff were appointed, and it is they who became responsible for translating the legislation into a planning framework. The Pinelands planning process was guided chiefly by two forces: (1) the legal requirements of the New Jersey Pinelands Protection Act (N.J. Senate 1979), which, in turn, was based on the 1978 federal act (U.S. Congress 1978), and (2) the personalities and attitudes of the Pinelands commissioners and staff.

Both federal and state acts distinguished between goals for the core, or preservation, and the larger buffer, or protection areas. Legislative goals for the preservation area clearly stated that the commission was "to preserve an extensive and contiguous area of land in its natural state," and "prohibit any construction of development which is incompatible with the preservation of this unique area" (sec. 8c). There was and still is little disagreement about the strict preservationist plan for this area, although several cranberry growers continue to grouse about state control, and some environmentalists feel that no development, no matter how small, should be allowed. Goals for the protection area, however, were more complex:

> The goals of the comprehensive management plan with respect to the protection area shall be to
> (1) Preserve and maintain the essential character of the existing pinelands environment, including the plant and animal species indigenous thereto;
> (2) Protect and maintain the quality of surface and ground waters;
> (3) Promote the continuation and expansion of agricultural and horticultural uses;
> (4) Discourage piecemeal and scattered development; and
> (5) Encourage appropriate patterns of compatible residential, commercial and industrial development, in or adjacent to areas already utilized for such purposes, in order to accommodate regional growth influences in an orderly way while protecting the pinelands environment from the individual and cumulative adverse impacts thereof. (sec. 8b)

The Pinelands Act also mandated specific areas of inquiry for the comprehensive management plan, which had to be adopted within eighteen months of the adoption of the act. The assessment had to include:

> [A determination of] the amount and type of human development and activity which the ecosystem of the pinelands area can sustain while still maintaining the overall ecological values thereof, with special reference to ground and surface water supply and quality; natural hazards, including fire; endangered, unique, and unusual plants and animals and biotic communities; ecological factors relating to the protection and enhancement of blueberry and cranberry production and other agricultural activity; air quality; and other appropriate considerations affecting the ecological integrity of the pinelands area; and . . . an assessment of scenic, aesthetic, cultural, open space, and outdoor recreation resources of the area. (sec. 7a)

The act granted the commission extraordinarily strong enforcement powers. First, the commission was responsible only to the governor, so it was separate from established state bureaucracies. Second, in almost all cases its powers superseded those of other state agencies in the region. Third, the commission assumed the home-rule police powers of local municipalities to plan and zone their counties and towns. The fifty-two towns and seven counties were given one year from adoption of the plan to "submit to the commission such revisions of the municipal master plan and local land use ordinances as may be necessary in order to implement the objectives of the comprehensive management plan and conform with the minimum standards contained therein" (sec. 11b). After conformance, towns and counties could then have planning and zoning powers returned to them, although final approval for all permits rested with a majority vote of the commission.

From the beginning it was clear the legislation posed three major problems. The first and worst was the ridiculously short time frame in which to create a plan and bring municipalities into conformance. The second was to create a plan that resolved conflicts inherent in the goals, and the third was to create a structure that would meld the efforts of federal, state, county, and local governments so the plan could be implemented. To solve these problems would take an expert, creative, and above all, empathetic staff and commission. Appointments to the commission and staff were, therefore, of the utmost importance.

Early in 1979 Governor Byrne appointed seven Pinelands commissioners, including the chairman, Franklin Parker, an attorney from the wealthy exurban county of Morris. Parker was a trustee of the New Jersey Conservation Foundation, an influential environmental group that concentrates its activities in the suburban and rural sections of central and northern New Jersey. Joining Parker was Candace Ashmun, then executive director of the Association of New Jersey Environmental Commissions; she had been active in New Jersey conservation concerns for many years, and her views on the Pines were similar to those of Byrne and Parker. Environmentalists in South Jersey, however, were suspicious of the NJCF, which, they complained, had done little of the hard infighting to pass Pinelands legislation on the federal and state level, and were latecomers to a battle already won.

Parker's and Ashmun's vision of the Pine Barrens was crucial in Pinelands planning because it dominated the commission's decisions, and they were politically well connected. That vision was of an untouched wilderness, a place that should be saved for recreation and scientific research for future generations, while some areas near the coast could be sacrificed to provide housing. In short, most of the Pines were to remain "pristine." Their understanding of local history, culture, and land-use patterns remained rudimentary. Governor Byrne also appointed two members of the Pinelands Coalition, Gary Patterson and Floyd West, both of whom represent preservationist viewpoints. Their vision was also of a pristine wilderness, and they viewed most Pinelands residents and politicians as enemies bent on destroying that wilderness. They wanted no further development of any sort in most of the region and wanted to treat most of the protection area in the same manner as the preservation area. Their opposites were appointed by the counties of Cape May, Ocean, and Atlantic, a minority who fought for home rule and a change in Pinelands legislation to dilute the power of the commission.

It quickly became clear that consensus on some of the most fundamental philosophical concepts would be impossible. Peacemakers did appear—Budd Chavooshian, the governor's appointee, a kindly and revered planner in the state and champion of the concept of transfer of development rights, and Robert Shinn, a Burlington County freeholder who was politically astute and in touch with major cranberry interests in his county. But the peacemakers could generally do little more than mask most underlying conflicts among commissioners. The different expectations of commissioners made communication difficult, and they could not focus on questions that would become critical to the plan's success or failure—intergovernmental cooperation, the sharing of power, and an accurate formulation of the desires of residents as well as outsiders. Those who expected a strict-preservation plan felt that several county commissioners were stooges for developers, while the latter saw some of the governor's appointees as silly, petty tyrants. They could not agree on a definition of the public interest, which one group saw as purely federal and the other as wholly local. The commission's lack of focus and direction was exacerbated by the lack of an executive director and a skeletal, interim staff between February and June.

In May, however, two consultants, Hal Williams and John La Rocca (1979) of the Institute on Man and Science in Rensselaerville, New York, wrote an insightful report to the staff and commission. The report relied on similar planning experiences in the Northeast and should have served as a set of guideposts for the commission over the next year and a half.

> The pieces are present here for a very exciting regional program which has national and even international implications as a model—bright people, thoughtful and provocative legislation, defined interest groups who have some impetus to cooperate, a good initial data base, etc. But, there is little evidence that the pieces will fit together well. Indeed, our sense is that things are not yet off to a good start. (1)

Williams and La Rocca presented to the commission a number of salient problems, which we include here in some detail because they apply to many regional-planning efforts and still haunt the Pinelands Commission in 1985.

> The path of least resistance is to define the primary interest groups of builder-developers and conservationist-ecologists as the public. Indeed the pattern for this is already set. . . . There are many people who invariably find it difficult to represent themselves effectively, including people with lower incomes and with very local orientations. Are they among the constituencies? A framework is needed to insure that those who will bear the consequences of decisions have some involvement in the process of making them.

> *We are strongly taken by the lack of a social ecology perspective.* Virtually all of the background work, legislative wording, and the discussions we have heard deal with conservation of physical features and economic impacts. A full range of social concerns must be addressed if the Pinelands Commission has any pretense of understanding the impacts of land use policies and programs on the quality of life of those who live in the Pinelands.
>
> Another reason for including a social perspective is that people as part of the landscape have *power* and *capacity*. An example is the sense of history and cultural heritage which we understand to be especially strong in the interior. This is a force whose inclusion can bring cultural and physical preservation *together* in an integral way which includes, binds, and commits proponents of each. . . . If the Commission does not spark a process which helps people to understand the benefits and pathways of change, "lifestyle" adjustments are likely to be few and far between.
>
> We see little thought yet given to how the regional plan can move from a posture of requiring conformance and constraint to one of enabling local initiative within clear limits. Unless the framework is made enabling, local units will be given neither the opportunity nor the responsibility of translating localism from reactive protectionism to initiative.

We do not have a sense that all Commissioners or interim staff know the Pinelands well. While this is more obviously true of those who do not live there, it is about as likely to apply to those who do. . . . Virtually no one person could possibly know the whole landscape unless [he has] taken a systematic effort to study it carefully.

Approach to public involvement needs considerable thought. The standard fare is the public meeting which was the format of the hearing we attended. The problems with this format are legion including these: They tend to bring out advocates and highly vested interests. They tend to favor people who are articulate and have more mobility. They often do not have a clear purpose. (It is announced that decision makers will take what they hear ''into account.'' But the way in which they will do so is rarely defined. They tend to make the citizen's role very passive.)

We see little evidence that experience to date has been documented, let alone analyzed. Documenting experience in order to make course corrections is not a luxury; it is a *necessity*. While some people feel that merely having experiences is enough to probe them, we have seen too many evidences to the contrary. (Williams and La Rocca 1979, 2–6)

Ten days after the Williams and La Rocca report was written, Governor Byrne signed the Pinelands Protection Act and appointed Terry Moore executive director. Moore was a politically astute man who had had great success running the Newark Watershed Company, a nonprofit organization that regulated watershed and reservoir lands in northern New Jersey for Newark's public water supply. Given Byrne's penchant for secrecy and his need to appoint people close to him and his staff, he made reasonable choices in Parker and Moore. Byrne was comfortable with their styles and agreed with their views of the Pine Barrens as a pristine wilderness about to be despoiled. The governor and his appointees had good intentions, but a misplaced sense of the way people and place worked in the Pines and a very poor understanding of the importance of public involvement in the planning process. The warnings of Williams and La Rocca went unheard.

From July 1979 through July 1980 the commission and its staff worked desperately to complete a plan. Even the most empathetic and brilliant staff would have had the utmost difficulty given the time constraints, and most of the young and energetic but inexperienced staff were over their heads. Furthermore, outside consultants could not be hired until September 1979 because the money to hire them took that long to come through.

What planning process occurred during that time? There is as yet no official Pinelands Commission document to explain what happened, but from our perspective the plan was born of confusion. We were hired by the commission in November 1979 to complete the historical and sociocultural assessments of the Pine Barrens and to work closely with staff on public involvement. Within a short time, it became clear that staff and commissioners were awash in a sea of consultant reports without a framework into which they could cast the data. The work plan called for information to be gathered, but there were no clear objectives and no method by which to incorporate data into a planning framework. In fact the consultants most responsible for the plan were not hired until December 1979, nor did the commission or executive director hire an assistant director to coordinate planning until November. The only section of the agency that seemed to be on target, although badly overloaded, was development review, the section that granted permits for development. In other words, by the end of 1979, data were being collected, development applications were being reviewed, but no planning framework had been established, no data-retrieval system was in place, and the public-involvement effort was in disarray.

In the frantic effort just to keep abreast of paperwork, the relationship between goals and implementation had been lost. Neither the staff nor the commissioners were predisposed to focus on public involvement, historical perspectives, or the human ecology of the Pines. The commissioners continued to argue bitterly over state versus local control. The coalition of commissions that formed around Franklin Parker and the governor's appointees left several county appointees and many local politicians bitter. Intransigence began to feed on itself, although much of the underlying resentment was masked by the celebratory atmosphere of the intense effort.

The formulation of the planning framework finally began when the commission hired the legal firm of Ross, Hardis, O'Keefe, Babcock, and Parsons (1980) and the planning firm of Rogers, Golden, and Halpern in the winter of 1979–80. These consultants, with three staff members, were chiefly responsible for pulling together material from the consultants' reports and creating a plan. The mandate to the planners and lawyers in the end amounted to this: Preserve as much land as possible for development, force as much development as possible into areas where development already exists, and make the plan legally impregnable. Given the short time frame and poor start of the process, perhaps this is as much as should have been expected. Expectations were high, however, and the results could not match unrealistic hopes for a plan that could respond to broad legislation. The commission could not resist using its police powers, the direct result of which was the power struggle between the commission and local governments.

Like the designers of the Adirondack plan, the planners established criteria for a variety of uses in different areas. Briefly, the criteria were based principally on physical and biological factors, and the components used to define ''the essential character of the Pinelands environment'' neglected to mention human activity. The commission then established criteria for designating forest (low-density-use) areas, agricultural areas, rural-development and regional-growth areas, and villages and towns. In each area certain uses were permitted and others prohibited, and development densities were given for all areas. The plan also allowed for Budd Chavooshian's pet concept called transfer of development rights, or TDR—a complex idea that allows a landowner to sell the right to develop a piece of property in an undevelopable area to someone else who is encouraged to develop housing in another area.

The counties and towns were responsible for bringing their master plans and zoning ordinances into conformance with the plan. As incentives, the plan offered some funding to those towns that complied, and the commission promised to pursue legislation to compensate towns for loss of tax revenue due to low property assessments. The plan also offered to request legislation to provide relief to landowners deprived of their rights to develop land. In addition the Comprehensive Management Plan provided for a series of programs to manage water, soil, vegetation, wildlife, forestry, cultural, and scenic resources. The draft plan was completed in June 1980, and the following November the final parts of the plan were adopted.

In the end, what can one say about the Pinelands Comprehensive Management Plan? From the viewpoint of the original commissioners and the staff, it is a resounding success. In his first-year review of the plan, Parker wrote in a memorandum that the commission was well on its way toward success, which he measured chiefly in terms of the number of towns conforming to the plan.

> Today, one year after the Comprehensive Management Plan took effect, the Commission is actively working toward conformance with nearly all of the 52 Pinelands Area communities. . . . I believe that we have shown on every occasion that the Pinelands plan is flexible and that the Pinelands Commission can satisfy local needs within the Plan's general guidelines. . . . Only in those municipalities which have deliberately ignored the Act and the Plan has the Commission been compelled to take steps to ensure the integrity of the Pinelands Plan. Local conformance is the key to the success of the Comprehensive Management Plan.

Given Parker's vision in 1979 and that of the majority of the commissioners and staff, the commission has indeed been successful. First, suburbanization in most of the Pine Barrens has been halted, and much of what outsiders view as the pristine environment has been preserved. Second, that pristine view of the Pines has been translated into the plan which expresses the aim of the Pinelands Act, "to preserve the essential character of the Pinelands environment." Third, the full power of the law has been made to stick legally by winning a series of lawsuits brought against the commission over the last three years. Last, a self-fulfilling prophecy has been created that has reinforced their vision of local people as recalcitrant and oriented toward development. One often hears from staff and commissioners, "We gave the locals every chance to express themselves, and if they don't want to obey the law, we'll have to enforce it."

From the viewpoints of some of the planners and lawyers, the jury is still out, and it will not be possible to assess the success or failure of the plan for several years. As one of the lawyers said in conversation at the end of 1982, "We could not build a magic document which would last forever. We could only achieve a certain percentage of our overall goals." From his perspective and that of several staff members the expectations of local people had to be changed. Towns that expected development had to be told it was not to be, and towns that wanted to limit development had to accept more than they had expected. In other words, towns had to be forced to comply very quickly. Their spirit had to be broken, and once broken, they could regain most of their home-rule privileges. The planners and lawyers knew some towns would accept the plan, some would bridle, and others would refuse; the first group would be rewarded highly, the second rewarded a little, and the last, punished. After the towns were brought under control, the rest of the planning process could begin, but three years after the plan's adoption, 80 percent of the staff's time is still occupied with permits, regulations, and conformance; staff and commission admit that this will be the case for the foreseeable future.

From the viewpoint of many residents, the plan is a failure. Although many insiders would like to preserve what they perceive as the essential character of the Pines, hatred of the commission has increased. Many towns conformed angrily, and several large ones in the middle of the protection area, as well as two counties, refused for three years to engage in any discussions. There are many horror stories to tell—people cheat to get what they consider a fair price for their lands; sections of the Pinelands, like the Forked River Mountains, are threatened by sand- and gravel-mining operations; enforced suburbanization threatens to destroy the character and economy of some towns; and the plan allows unsound environmental practices in towns that have not conformed to the plan.

In fact, it is difficult to make a final judgment at this point. The plan has clearly prevented development in some areas that should not have been developed, but it has serious shortcomings. We can state, however, that despite time constraints and other difficulties in preparing the plan, it could have been much better. Such a judgment is not even a matter of hindsight because the warnings were clearly expressed before the planning process even began. Just as one cannot make good wine from bad grapes, one cannot make a good plan from ill-conceived premises.

The sources of the Pinelands commissioners' and staff's misconceptions are common to such planning attempts. In a recent dissertation on a small community in northern California, John F. Salter (1981, 7) succinctly suggested that:

> If there are any villains in conflicts between the community and outside agencies such as the [Pinelands Commission], they are not so much individuals or groups, as the monothematic and solely purposive quality of today's development and planning. Regardless of their ultimate goals, such plans are frequently conceived in isolation and applied without a textured appreciation of local contexts and consequences.

Salter here uses the term purposive in the sense of single-minded adherence to a goal, such as ''preservation,'' rather than an approach grounded in an understanding of the human ecological system as a series of adaptive strategies. The basic struggle, whether in northern California or southern New Jersey, has to do with the methods one uses to deal with whole systems. Salter continues:

> One [method] tends to be flexible, organic, rich in adaptive potential; the other tends toward rigidity and simplified linear frames of analysis. This latter, analytic model is dominated by a short-term purposive consciousness and lacks the capacity for successful, that is to say, sustained adaptation. Within small communities adaptation has traditionally required the ability to recognize subtle shifts in interpersonal relations, and further, to test endlessly the appropriateness of one's actions in relation to the various contexts within which the resident of the small community exists. This contrasts with the professional training regimens of mainstream institutions [such as the Pinelands Commission], which either downplay the significance of contextual awareness, or deal with context-related complexities through a variety of persuasive or manipulative techniques. Ignoring or misinterpreting local culture is a frequent source of conflict between Shadow Forks [or Pinelands] residents and these institutions. (9)

Much regional planning in America simply does not deal adequately with the nature of complex socionatural systems, a problem we take up in more detail in the

Appendix. The major reason for this lack, which leads more directly to obfuscation than to revelation, is the tendency for planners to concentrate on "issues" without regard to their context and in ways similar to crisis management in business or government, issues such as: development versus preservation, least-cost housing versus redevelopment for condominiums, farming versus suburban development, and a host of others. Techniques, which planners refer to as tools, and which can be anything from zoning to tax incentives, are then used to address the issues. Because it is so issue oriented, planning tends to deal in the immediate present without benefit of the kind of framework we need to understand the impacts of planning decisions. To give just one example from the Pinelands process: the lack of a suitable framework, the hiring of an inexperienced staff, the limited vision of most commissioners and a ridiculously short time frame led to oversimplification of the planning problems facing the commission. The easiest way to develop a plan was to focus on the development-versus-preservation issue and the use of tools to encourage development in some areas and prevent it in others. Thus, to many commissioners and staff, the real issue became boundary lines—zoning districts—delineating where certain, very specific uses could and could not occur. In answer to the question, "Could you not have made a more responsive, less restrictive plan?" the usual answer is, "Well, someone has to draw the lines." That narrow view of planning is pervasive. Not only does it beg the question of what planning should be about, but it so confuses goals, issues, methods, and tools that in the end the tool itself (zoning in this case) becomes the central core of the planning effort. The confusion is self-perpetuating, and those who become most adept with the use of tools often become the most highly respected and highly rewarded practitioners.

In fact the real issue in much of planning is to enact the goals of a plan while maintaining the health of a region which, in turn, rests on the ability of its human ecology to adapt to changing conditions. That ability is the subject of the main body of this work which described the processes by which people and their environment work in flexible and interactive systems. Planners must not underestimate the role flexibility plays in the workings of a place. One of the primary functions of the planners' task is to reveal its special nature. Our concept of flexibility, incidentally, should not be confused with the term as it's often misused in official circles: to refer to political trade-offs. Pinelands officials, for example, claim they are being flexible when they change municipal boundary designations for various uses. Again we refer to Salter's (1981, 22–23) study:

> Flexibility entails a series of more or less closely related processes. These include balance, limits, and the ability to correct oscillations in the system which, if uncorrected, would tend to go out of control through processes of positive feedback. Bateson has defined flexibility as "uncommitted potential for change" (1972, 497). The positive survival value of flexibility lies in this characteristic of not being locked, perhaps fatally, into a predetermined or uncorrected sequence of development.
>
> In contrast to flexibility, prejudice or prejudgment foreshortens and eliminates the qualities which characterize successful adaptation. It is in this sense that Kant characterized prejudice as a form of judgment which refuses to correct itself. And although it is true that a biological or social system may in the short run be dominated without consideration for flexibility, there is an increasing awareness

that to do so indefinitely is possible only at the cost of the system itself. While prejudice is the inability to correct judgments at a variety of levels, racial, cultural, economic, philosophical, and epistemological, it is from another perspective the denial of flexibility, the inability to cross thresholds, to move from context to context, to linger between contexts and to keep in mind that there are a variety of thresholds to be considered in any social or biological system. In contrast to prejudice, which is a mechanical response, flexibility invokes a sense of self-correcting organic tendencies which are co-adaptive as opposed to being merely dominating.

The pleas of ourselves and others for more responsive planning, however, should be mistaken neither for naïve yearnings to return to the old ways nor advocacy for pure local control of government. The locals are not always right, or the planners always wrong, but to be consistently in the grip of a locals versus planning-officials struggle is to admit that planning, as it is often practiced, has serious problems. It may be too late to change the Pinelands planning process because so little communication occurs, visions of the future are rigidly set, and a fortress mentality surrounds most Pinelands commissioners and staff. For example, three years after adoption of the CMP, the Pinelands Commission is reviewing the plan's impact, and the commission, even before the review process is established, has decided to make no major changes in the CMP. The thrust of Pinelands planning efforts has been focused on defense of the plan. It is a form of judgment which refuses to correct itself, and, indeed, one can find nowhere in the minutes of the commission's meetings acknowledgment that mistakes have been made. To admit mistakes, the commission feels, is to leave itself open to a judicial suit which, should the suit be lost, would defeat what has become the purpose of the CMP—to defend itself.

This bind is by no means limited to planning in the Pine Barrens. It is an unfortunate result of the way planners often do business in the contemporary world, but it is not a necessary result. What we offer in the rest of this chapter are suggestions to help us out of these binds. We do not intend to restructure the whole of planning, but if we come to our first task of revelation with a different frame of reference, ready to listen as well as to explain, we will have already won part of the battle.

Other people have listened well in more difficult regions of the world. In 1974, Judge Thomas Berger of the city of Vancouver, British Columbia, began an inquiry into the terms and conditions that should be imposed with respect to any right-of-way granted across Crown lands for a proposed Mackenzie Valley pipeline, one which was in fact never built. Judge Berger's letter to the Minister of Indian Affairs and Northern Development (1977, vii) is an example of what a good listener can hear in a true wilderness area:

> The North is a frontier but it is a homeland too, the homeland of the Dene, Inuit, and Metis, as it is also the home of the white people who live there. And it is a heritage, a unique environment that we are called upon to preserve for all Canadians.
>
> At the formal hearings of the Inquiry in Yellowknife, I heard the evidence of 300 experts on northern conditions, northern environment and northern peoples. But, sitting in a hearing room in Yellowknife, it is easy to forget the real extent of

the North. The Mackenzie Valley and the Western Arctic is a vast land where people of four races live, speaking seven different languages. To hear what they had to say, I took the Inquiry to 35 communities—from Sachs Harbour to Fort Smith from Old Crow to Fort Franklin—to every city and town, village and settlement in the Mackenzie Valley and the Western Arctic. I listened to the evidence of almost one thousand northerners.

The Canadian wilderness is a far more inaccessible and more difficult place to interpret. We can expect no less an effort from our officials than that given by Judge Berger.

A Humanistic Approach

What are the people of *Water, Earth, and Fire* saying to the professional planner? They are saying, "We are part of the unique character of the Barrens; our way of life, our land-use patterns, our concerns are just as valuable as the unique vegetation, the rare tree frogs, and the historic sites and roads." They are saying, "We want to have a role in the planning and management of this region." They are saying, "We have the ideas, talent, skills, and expertise that you need to manage the Pines." These are not new demands placed upon the planning profession, for there is a trail of articles and books that goes back to the English city planner Sir Patrick Geddes (Stalley 1972) who called for anticipatory research, or, in Geddes's words, "a diagnostic civic survey" that matches a specific planning goal with a particular social or cultural setting.

The Pinelands Commission had the benefit of anticipatory research that detailed the various agendas, both local and outside, with respect to the planning and management of the Pinelands Reserve, but staff members found it irrelevant to pursuance of the planning goals. Thus the purpose of this section is to translate such research into five propositions that can help resource managers integrate socionatural information into the published sets of goals, standards, and programs that make up the formal planning process.

Again we emphasize that the propositions are guidelines for integrating flexibility and understanding into regional planning, not a panacea for conflicts inherent in planning processes. For example, whenever a state legislature passes a law, such as the Pinelands Act, to regulate or prohibit development in an area, people inside that area will see some of their property values decline, while many people adjacent to the preserved area will find increases in value. This common problem is called the windfall-wipeout syndrome. It is, of course, much more complicated than who gains or loses economically; some owners may prefer lower values in return for privacy and protection, and landholders in either the windfall or the wipeout zone may have different interests depending on their economic situation, age, stage of life, disposition of their spouses, or whether they live there or are absentee landowners. In any case, no set of propositions or guidelines can obviate the windfall-wipeout problem or inherent generational, socioeconomic, psychological, or cultural conflicts.

The following propositions, however, can help alleviate, and in some cases dissipate, the anger and long-term resentment, especially on the part of local residents, that attend any planning process. We listened to the people of the Pine Barrens, and we have tried to translate their concerns into planning procedures. Our anticipatory research into the themes of love of place, the collective memory and the seasonal cycle, point and counterpoint, and the changing balance between family, community, resource availability, technology, and markets allowed us to understand in context the everyday conversation and concerns of residents, users, and managers. We call our procedures ''propositions'' because they are an argument for and a demonstration of the way in which the planner's prescriptive overview can take into account, and indeed be enriched by, the user's intimate and pragmatic knowledge of place. The propositions state the goals of sound planning and suggest procedures for implementing them.

(1) Maintain regional control but decentralize the planning process through relevant public participation. Base the designation of subregional areas for local planning on contemporary and historical patterns of use.

(2) Select when possible management strategies best adapted to the social, economic, and ecological arrangements in a subregion. A knowledge of the balance among family, technology, resources, and markets provides the necessary insights.

(3) Recognize landscape patterns of use and tradition as a basis for siting new uses. Landscape analysis based on the themes of love of place and the switching seasonal economy provides the necessary interpretation for these siting guidelines.

(4) Incorporate local skills and actors into site management. Mix these skills and ideas with applied science. Anticipatory research on the seasonal cycle and the collective memory reveals these skills and patterns.

(5) Recognize local and subregional aesthetic norms in site design and management. Data from each theme contribute to this understanding.

Incorporation of these procedures into the standard land-use-planning process will not only protect the environment, but enhance the quality of community life. The geographer Meinig (1972) has shown that local cultural values are found simultaneously with local environmental values. Our propositions should help tap this ''new consciousness of local, cultural, and environmental values.'' Throughout the rest of this chapter we will give examples of each proposition applied to the Barrens as a whole or to a subregion. We will emphasize the synthesis between the planner/scientist view of the region and the user's view of place, and we hope, in comparing our propositions to the requirements of the Pinelands plan, that the use of socionatural concepts will prove beneficial.

PROPOSITION 1

Maintain regional control, but decentralize the planning process through relevant public participation. Base the designation of subregional areas for local planning on historical and contemporary patterns of use.

How can a plan achieve its legislative goals throughout a region? The response of the Pinelands Commission was to establish a series of use zones on top of existing land-use patterns in the Pinelands (maps 8, 9). Standard acreages and permitted uses were established for forest, agricultural, rural-development, and village areas, and densities were established for each township and regional growth area. The result was that the Pinelands Commission developed more a zoning ordinance than a master plan, and most of the commission's and the staff's time subsequent to the plan's adoption has been spent on enforcement, regulation, and conformance, not on planning and resource

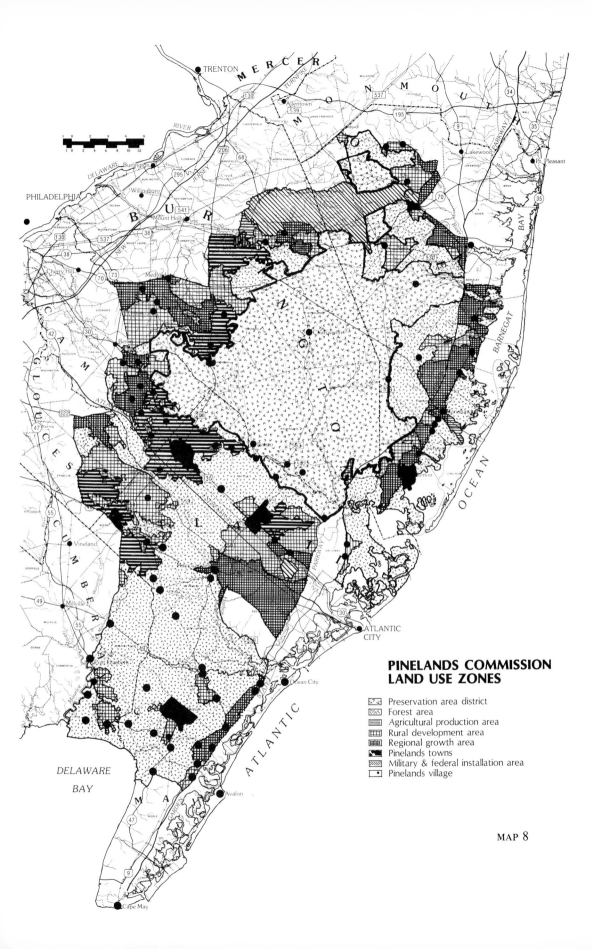

PINELANDS COMMISSION
LAND USE ZONES

- ⬚ Preservation area district
- 〜 Forest area
- ▤ Agricultural production area
- ▦ Rural development area
- ▦ Regional growth area
- ◆ Pinelands towns
- ▨ Military & federal installation area
- ⊡ Pinelands village

MAP 8

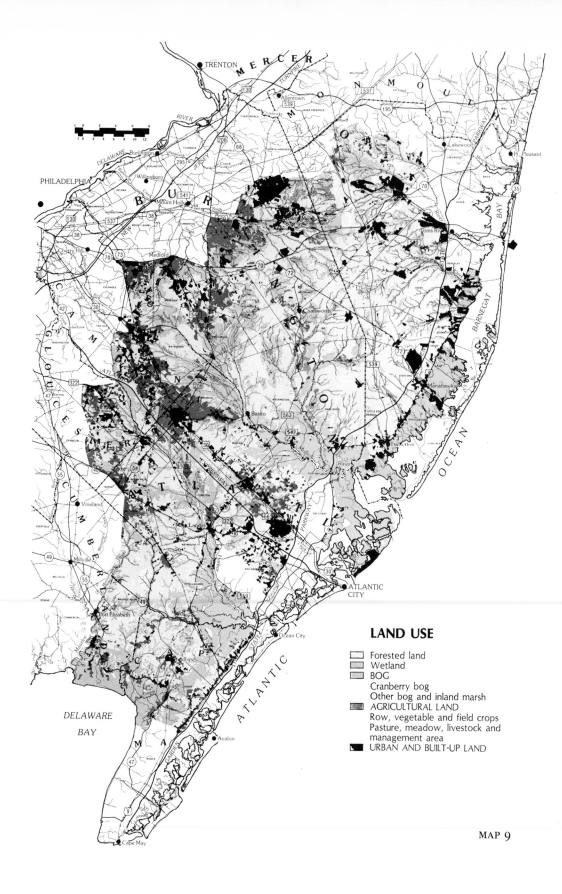

LAND USE

Forested land
Wetland
BOG
Cranberry bog
Other bog and inland marsh
AGRICULTURAL LAND
Row, vegetable and field crops
Pasture, meadow, livestock and
management area
URBAN AND BUILT-UP LAND

MAP 9

management. The CMP did reflect the visions of many legislators and people who live outside the Pine Barrens. Of the seven components used to delineate forest areas, only one—cranberry cultivation—involved human activities, and these same seven components were used to describe ''the essential character of the Pinelands,'' to use the wording of the Pinelands Protection Act. As we have seen, the vision of a Pine Barrens without people corresponded to that of many environmentalists, commissioners, and staff. In fact, each zone corresponded to some aspect of the Pinelands Act and responded to some interest group. There were also agricultural zones for farmers and regional growth zones for developers. The CMP even included a so-call Piney exemption whereby, in the preservation district, only a person who demonstrated ''cultural linkage'' could build a house. To qualify for cultural linkage a person must be ''a member of a two-generation extended family that has resided in the Pinelands for at least twenty years'' (sec. 5-302:A[1][b]).

No one can claim that the CMP did not embody the values that most New Jersey legislators had in mind: The courts have upheld the CMP in legal challenges, and the legislature has refused steadfastly to consider amendments to the act. But the plan does not correspond to the socionatural realities of different sections of the Pines. Some examples: The arbitrary half-mile boundary around each village is too large for some and too small for others; the flat ten-acre-per-dwelling-unit agricultural zone does not correspond to most farming practices; the permission to mine sand and gravel in many areas threatens the ecological integrity of parts of Ocean County; the flat densities in many regional growth areas threaten the cultural and economic integrity of some townships. We will see some of these problems in more detail throughout the rest of this chapter.

Is there a better way? We suggested in chapter 1 that the use of socionatural data allows us to obtain the kind of detail needed to make more careful decisions and respond to legislative goals more accurately and flexibly. Nowhere is this more useful than in our suggestions to delineate subregions. While recognizing local variations in human activities and ecological parameters, the commission, state legislature, and Congress will continue to maintain control of planning functions and do it with a more accurate understanding of the essential character of the region. In so doing, they can obtain maximum feasible public participation as well.

We expect a regional commission to promulgate environmental protection performance requirements for all current and prospective uses, but such requirements should not be arbitrarily established. Water-quality standards are a good example: The Pinelands Commission established boundaries of its forest zones, which constitute more than half the protection area, chiefly on the basis of ''pristine'' water quality. Undisturbed watersheds in the Pine Barrens have a nitrate concentration of 0.17 parts per million, less than rainwater with an average of 0.2 ppm. The commission used the pristine standard to establish an average of 17 dwelling units per square mile throughout the forest zones, both upland and wetland. That average would maintain pristine standards based on the impact of human waste from septic systems. To give readers a basis for comparison, drinking-water standards in New Jersey are 2.0 ppm nitrates.

Readers who understand the Pine Barrens' diversity, however, are aware that an average density of one house for every 39 acres in forest zones does not make sense in a socionatural context. Yet we must still find an alternative to such zones that will survive court tests, respond to legislative mandates, and reflect local conditions.

We recommend that the planning process begin not with a zoning map, but with the designation of policy regions based on the patches of the mosaic, the socionatural systems. An analysis of the relationships among policy regions can lead to a clustering of them into subregions. The regional authority could begin the process of plan implementation and use its mandate for public participation not to present previously synthesized conclusions for the entire million-acre reserve, but to devise subregional land-use control plans. Minimum standards for each subregion can be established based on both legal and socionatural requirements. While conflicts among various economic or social interests will not be avoided, each region will relate more accurately and flexibly to legislative goals, and the sum of the policies set for each subregion will then achieve the goals of the plan. There will be less tendency to make outdoor museums of places where development is prohibited or to overburden economic foundations and destroy the cultural coherence of places required to accept fast development.

We used a map-overlay method to delineate the policy regions of the Pine Barrens and relied on the distinctive patterns of the abiotic and biotic environments of the outer coastal plain as the first building blocks in our process. Upon these we sequentially examined various periods of historical land use. Finally, we dealt with the contemporary period through a map of potential residential development, a map of the areas of concern for a full range of local and regional land-use and social voluntary associations, an assessment of the type of local political structure, and a map of land- and water-use areas critical to the subregional economies. We interpreted the concurrences of these factors into regions based on the local patterns of land and water use.

Delineation of Subregions.   The delineation process to define subregions is not dissimilar from that of McHarg's (1969) in that we use an overlay system to find where congruencies and dissimilarities exist. However, we add historic and cultural information to the basic complex of ecological material, thus creating a socionatural basis on which to define subregions. The Pine Barrens are a result of the concurrence of climate, geology, hydrology, soils, and biota with three hundred years of localized European and American land use and a range of state, regional, and national forces that help shape the politics, economics, and culture of the region. As we proceed through this complicated material, however, patterns will quickly begin to emerge.

A map of surficial geology (map 10a) distinguishes regional variations in surface materials (map 10b), and such differences will remain distinct on contemporary vegetation and land-use maps (see maps 7a, 9). A map of the depth to seasonal high-water table (map 10c; see key to map 3), derived from soils information, adds a second mosaic of patterns that confirms the general geologic regions and simultaneously enriches our understanding by indicating local variations (map 10d).

To the first four maps we add the watershed divides (map 10e), which resulted from a combination of surficial geology and the downward cutting of river channels (map 10f). Regional and local land-use differences have historically been linked to the settlement and resource exploitation of different watersheds. The map of historic land-use patterns in the last quarter of the nineteenth century (map 10g) bears a striking resemblance to settlement and land use in different watersheds today. Each historic period added and subtracted patterns. For example, cedar swamps became iron mines that became cranberry bogs that became lake communities.

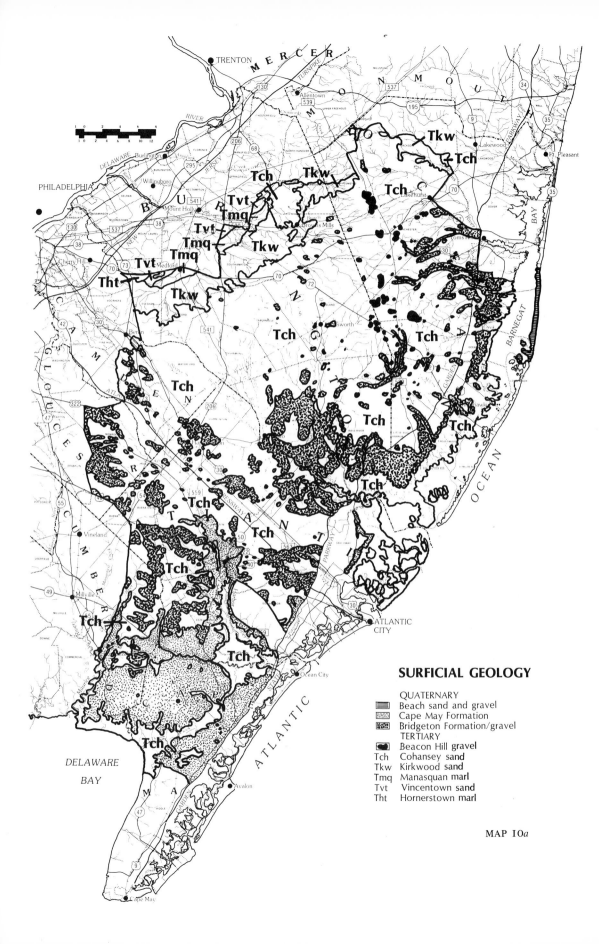

SURFICIAL GEOLOGY

QUATERNARY
- Beach sand and gravel
- Cape May Formation
- Bridgeton Formation/gravel

TERTIARY
- Beacon Hill gravel
- Tch Cohansey sand
- Tkw Kirkwood sand
- Tmq Manasquan marl
- Tvt   Vincentown sand
- Tht   Hornerstown marl

MAP 10a

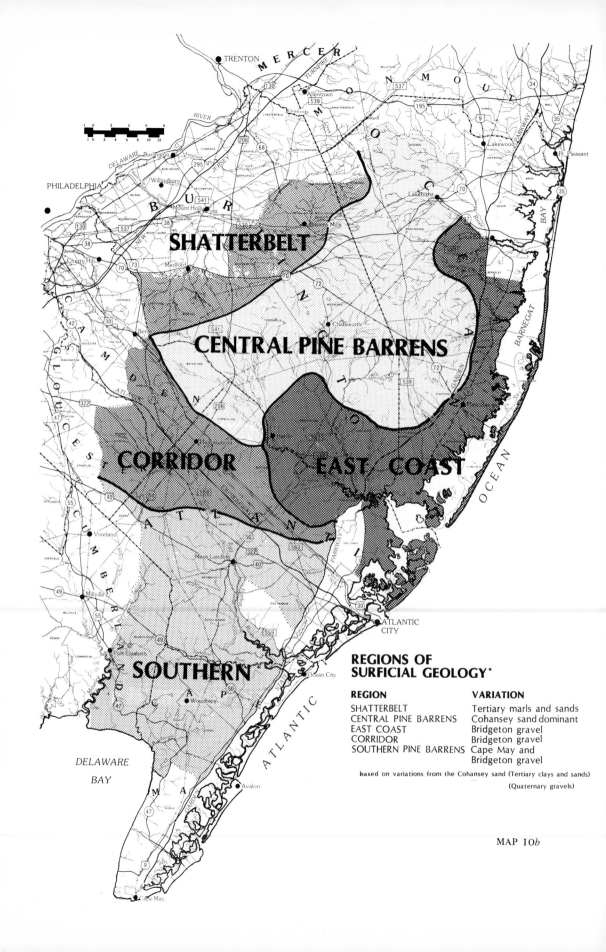

SHATTERBELT

CENTRAL PINE BARRENS

CORRIDOR

EAST COAST

SOUTHERN

REGIONS OF
SURFICIAL GEOLOGY*

REGION	VARIATION
SHATTERBELT	Tertiary marls and sands
CENTRAL PINE BARRENS	Cohansey sand dominant
EAST COAST	Bridgeton gravel
CORRIDOR	Bridgeton gravel
SOUTHERN PINE BARRENS	Cape May and Bridgeton gravel

based on variations from the Cohansey sand (Tertiary clays and sands)

(Quaternary gravels)

MAP 10*b*

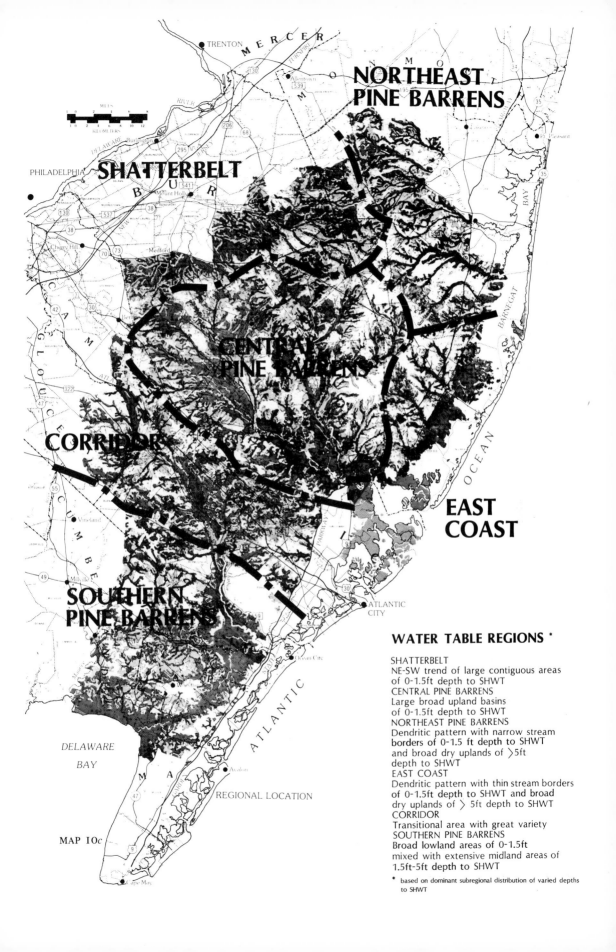

NORTHEAST PINE BARRENS

SHATTERBELT

CENTRAL PINE BARRENS

CORRIDOR

EAST COAST

SOUTHERN PINE BARRENS

PHILADELPHIA

TRENTON

DELAWARE BAY

ATLANTIC

REGIONAL LOCATION

MAP 10c

WATER TABLE REGIONS *

SHATTERBELT
NE-SW trend of large contiguous areas
of 0-1.5ft depth to SHWT
CENTRAL PINE BARRENS
Large broad upland basins
of 0-1.5ft depth to SHWT
NORTHEAST PINE BARRENS
Dendritic pattern with narrow stream
borders of 0-1.5 ft depth to SHWT
and broad dry uplands of >5ft
depth to SHWT
EAST COAST
Dendritic pattern with thin stream borders
of 0-1.5ft depth to SHWT and broad
dry uplands of > 5ft depth to SHWT
CORRIDOR
Transitional area with great variety
SOUTHERN PINE BARRENS
Broad lowland areas of 0-1.5ft
mixed with extensive midland areas of
1.5ft-5ft depth to SHWT

* based on dominant subregional distribution of varied depths
  to SHWT

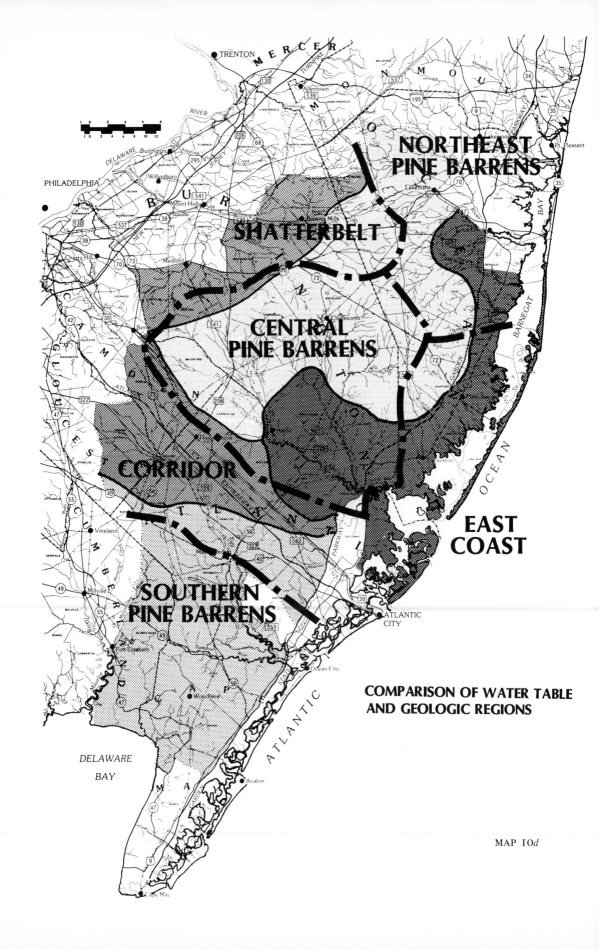

NORTHEAST
PINE BARRENS

SHATTERBELT

CENTRAL
PINE BARRENS

CORRIDOR

EAST
COAST

SOUTHERN
PINE BARRENS

COMPARISON OF WATER TABLE
AND GEOLOGIC REGIONS

DELAWARE
BAY

MAP 10d

We now have a rough regional analysis of distinct areas (map 10h). The coastal region, with its supportive upland, is subdivided by inlet and bay systems. There are distinct estuarine and tidal-river systems with their own settlement and land-use patterns and a large unbroken forest area north of the Mullica River bounded on the east by the coastal strip and on the west by croplands. Within this forest area are the watersheds of the large cranberry growers and the major state-owned forest and recreation lands. South of the Mullica lies another forested area, a diverse place with many types of vegetation and old industrial centers. Once it, too, supported cranberries, but today only blueberries and row crops are planted. Although it is diverse, it maintains its coherence from the Atlantic coast, across the uplands to the swamps and meadows of Delaware Bay and the fields and orchards to the northwest.

The western border of the Pine Barrens is the shatterbelt, a mixed area of inner and outer coastal-plains vegetation. Here are the major row-crop agricultural resources of the Pines, some surrounded by pines on the more sandy ground, some stretching unbroken to the west. Here are the border towns, commercial rather than woodland in their patterns and architecture, where the settlements resemble the Anglo-dominated regions to the west, not the architecture of the coast, the tidal towns, or the forest.

Of particular interest is the correlation between land-use patterns and watershed divisions (map 10i; see key to map 9). By analyzing contemporary land-use and social patterns in the Pines, we can break down the rough regional map more precisely into sociocultural regions (map 10j). After World War II, traditional cultures began to mix more quickly and widely with those of the larger Delaware Valley and New York regions. Year-round and vacation houses appeared in suburban patterns, and cultural and personal values began to mix and conflict with more frequency. One can now find a wide variety of places; some are dominated by gravel quarries, others are cranberry watersheds and state forests, and some have suburban/rural settings with a full range of volunteer social-service associations, realtors and business groups, environmentalists, and a variety of churches. Retirement communities dominate some sections, while others consist of traditional woodland hamlets with scattered settlement, a Methodist church, a fire company, a few hunt clubs, and agricultural lands that farmers still own rather than lease. An outline of these patterns reveals the many users who have a stake in the future of the Pine Barrens. There are forty such sections in the regions, and they add the dimension of local control to historical and contemporary cultural patterns.

The addition of county and township boundaries completes the process for delineation of policy regions (map 10k). The political subdivisions are essential because county and township taxing and police powers are affected by land-use controls. People will pay the costs or reap the benefits of political decisions township by township. With surprising regularity townships follow the basic watershed and vegetation patterns because historic and contemporary land-use patterns relied on the resources provided by the natural resources of the subregions. The reasons for clustering townships into subregions differ according to the socionatural characteristics of the place. Along the coast, for example, the following criteria are important: clam-lease lot patterns, upland watersheds, bay and inlet systems, development pressures, fire-control districts, and social and political organizations. In the northern forest region, one must look closely at cranberry and state ownership patterns, the layout of woodland villages, hunting and trapping territories, and the importance of fire. In the adjacent shatterbelt, one must consider development pressure, farmers' leasing patterns, voluntary associations, and local watersheds.

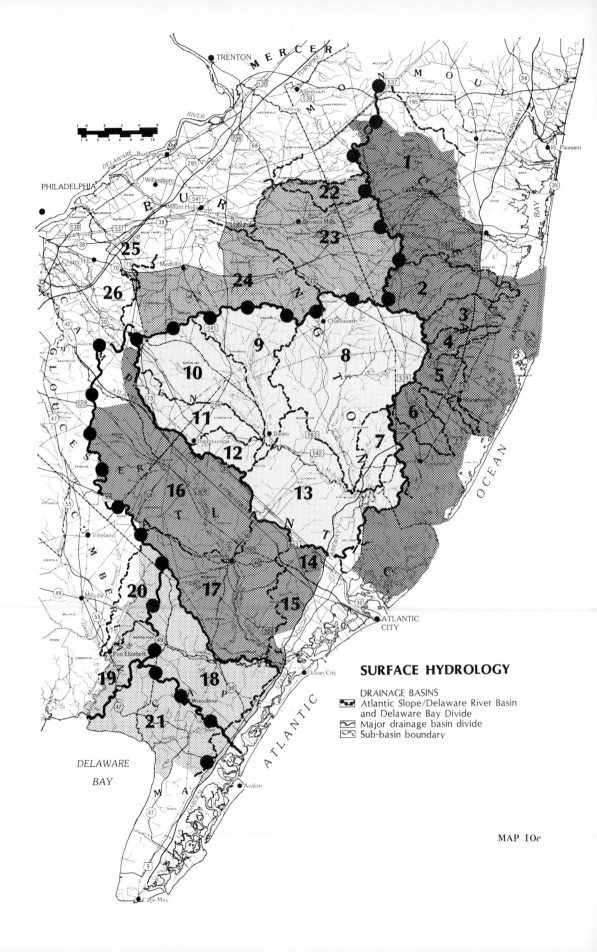

SURFACE HYDROLOGY

DRAINAGE BASINS

Atlantic Slope/Delaware River Basin and Delaware Bay Divide

Major drainage basin divide

Sub-basin boundary

MAP 10e

Within counties we grouped townships according to the dominant natural, histori-cal, cultural, and economic features in the subregion. Policy regions, therefore, encompass concurrences of water, earth, and fire linked by seasonal use, natural fluctuations, local social and political organization, the impact of development pres-sure, and local abilities to cope with change (map 10l). The map of projected urbanization (map 10m) confirms our subregional designations. Areas with little or no development pressure stand out as clearly as those under pressure.

The Pinelands policy regions are a synthesis of the cognized and operational models. Land users and local officials should agree with the designations because they match patterns of use and control. Scientists and policymakers, who take a regional view, should recognize in these policy regions the basic natural and cultural structure of the Pine Barrens.

During the actual planning process, the Pinelands Commission held two sets of meetings and one set of legally required hearings. In each case the entire plan was open for discussion, and no material was targeted to specific local regions. Two conclusions are possible: First, the geographic spread of meetings and hearings suggests that many areas were left out, thus affording little or no control or comment about specific regulations for the subregions; and second, the small number of meetings, all of which concentrated on the whole one million acres, made it difficult to create a plan responsive to local as well as regional needs. The Pinelands Commission chose a "top-down" planning process with participation used as a means of presenting material already synthesized and meeting state and federal regulations for public involvement.

Our suggestion to use the policy regions as loci for planning would allow for maximum participation and integration, but it would require time and patience. It would be a circular or iterative, rather than a linear, process, but it would be desirable and practical. In Walworth County, Wisconsin, during a planning process designed to develop and implement agricultural land-protection strategies, the local planning agency organized over five hundred meetings. The entire process took less than two years (Coughlin and Keene 1981).

We would suggest grouping the policy areas into subregions that could be used as focal points for decision making. It would be useful to have a representative group of

MAP-10e

		Subdrainage Basins		
1	Toms River		14	Absecon Creek
2	Cedare Creek		15	Patcong Creek
3	Forked River		16	Upper Great Egg Harbor River
4	Oyster Creek		17	Lower Great Egg Harbor River
5	Mill Creek		18	Tuckahoe River
6	Westcunk Creek		19	Maurice River
7	Bass River		20	Manumuskin River
8	Wading River		21	Dennis Creek
9	Batsto River		22	Crosswicks Creek
10	Atsion-Sleeper Branch		23	North Branch Rancocas Creek
11	Nescochague Creek		24	South Branch Rancocas Creek
12	Hammonton Creek		25	Pennsauken Creek
13	Lower Mullica River		26	Cooper River

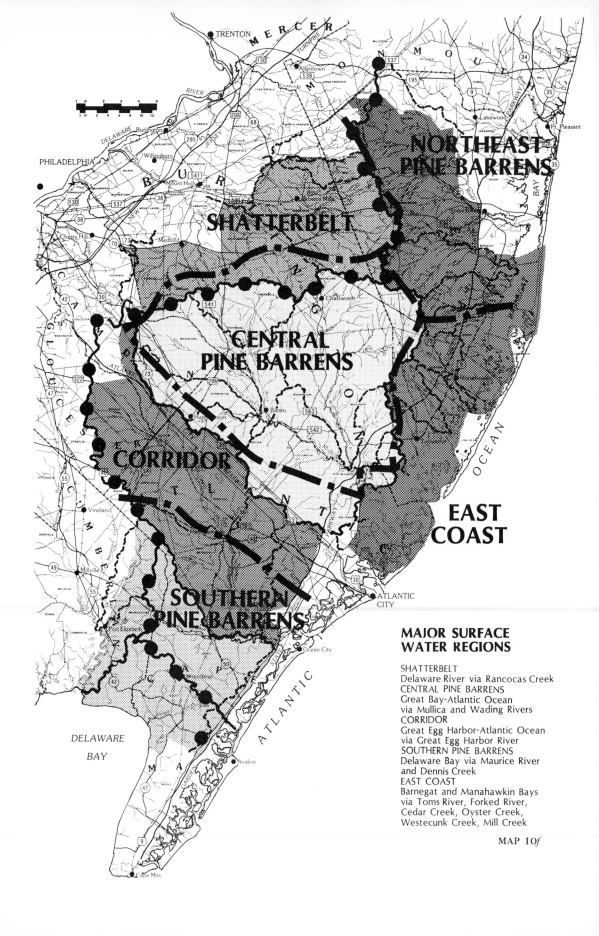

**MAJOR SURFACE
WATER REGIONS**

SHATTERBELT
Delaware River via Rancocas Creek
CENTRAL PINE BARRENS
Great Bay-Atlantic Ocean
via Mullica and Wading Rivers
CORRIDOR
Great Egg Harbor-Atlantic Ocean
via Great Egg Harbor River
SOUTHERN PINE BARRENS
Delaware Bay via Maurice River
and Dennis Creek
EAST COAST
Barnegat and Manahawkin Bays
via Toms River, Forked River,
Cedar Creek, Oyster Creek,
Westecunk Creek, Mill Creek

MAP 10*f*

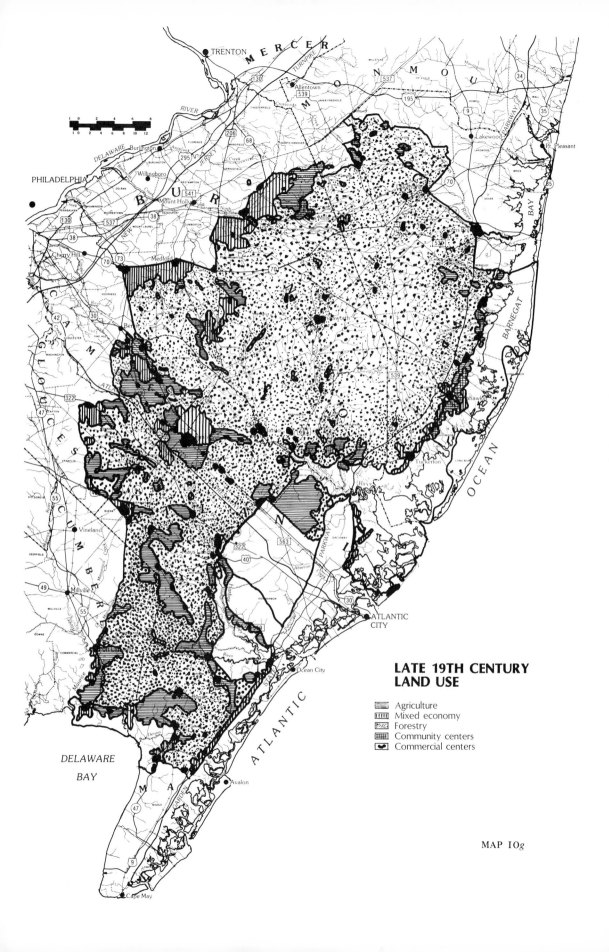

LATE 19TH CENTURY
LAND USE

▨ Agriculture
▥ Mixed economy
▨ Forestry
▤ Community centers
◡ Commercial centers

MAP 10g

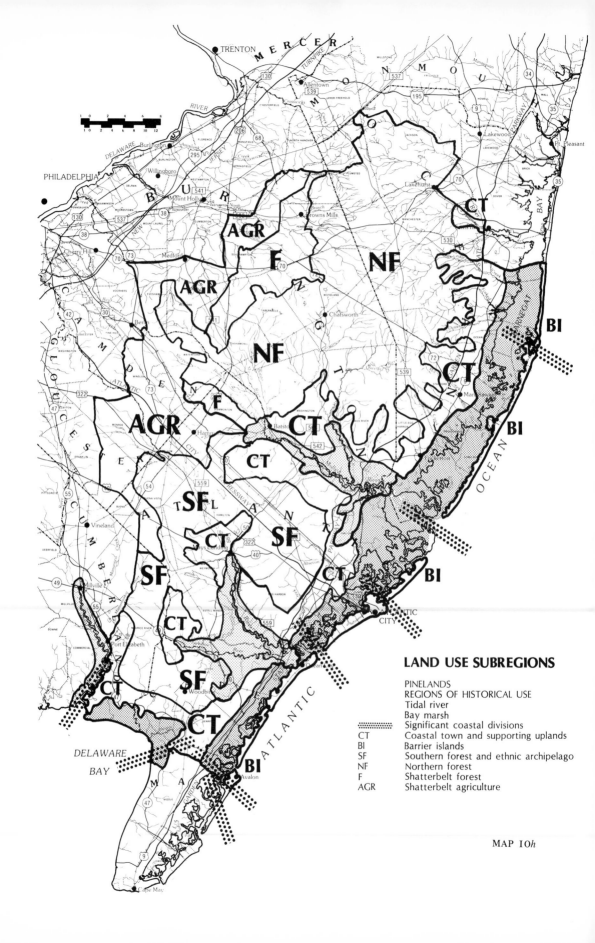

LAND USE SUBREGIONS

PINELANDS
REGIONS OF HISTORICAL USE
Tidal river
Bay marsh
:::::::::: Significant coastal divisions
CT   Coastal town and supporting uplands
BI   Barrier islands
SF   Southern forest and ethnic archipelago
NF   Northern forest
F   Shatterbelt forest
AGR   Shatterbelt agriculture

MAP 10*h*

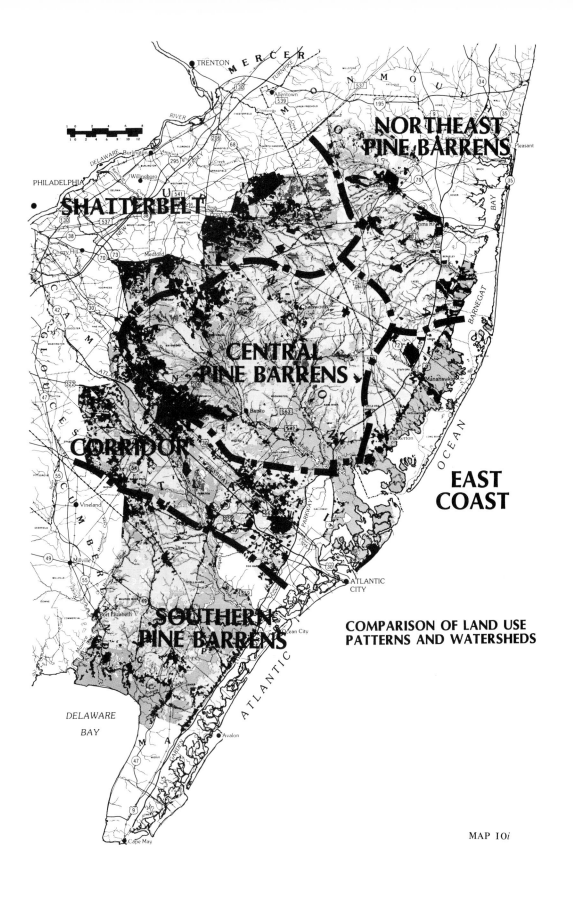

COMPARISON OF LAND USE
PATTERNS AND WATERSHEDS

MAP 10*i*

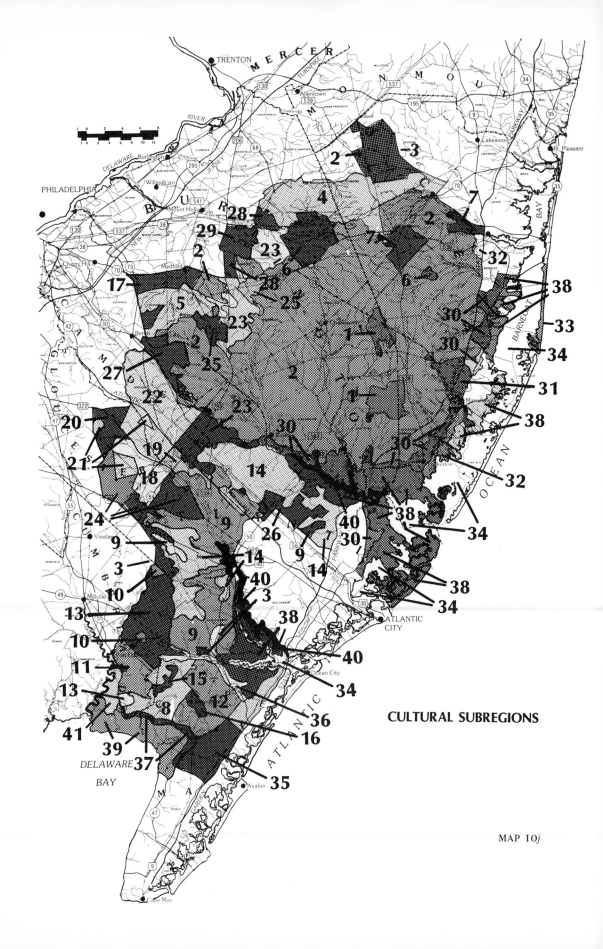

CULTURAL SUBREGIONS

MAP 10*j*

MAP-10j

	Northern and Southern Forest Region
1	The plains
2	North of the Mullica throughout Burlington and Ocean counties
3	Cranberry watersheds
4	Military
5	The lakes: Medford, Evesham
6	The lakes: Pemberton, Hampton, Bamber
7	Retirement: Manchester Township, Leisuretown
8	Belleplain State Forest
9	Uninhabited forest: Atlantic County
10	Uninhabited forest: Cumberland County (Land Scam Development)
11	Uninhabited forest: Cumberland County (sand and gravel)
12	Uninhabited forest: Cape May County (ex-agriculture woodland)
13	Inhabited forest: Cumberland County
14	Inhabited forest: Atlantic County
15	Primary use agriculture: Belleplain
16	Industrial / commercial: Woodbine

Western Crop Agriculture and Rural Suburban

17	Rural / residential: Medford, Marlton
18	The lakes: Monroe
19	Rural residential: Folsom
20	The "pikes": Williamstown, Berlin
21	Rural road settlements: Monroe
22	Rural residential / suburban: Winslow
23	Primary use agriculture: berries, vegetables, ethnic town center (Hammonton)
24	Primary use agriculture: Buena and Franklin field crops
25	Agriculture / forest transition: Medford, Shamong, Tabernacle

26	Primary use agriculture: ethnic town center (Egg Harbor City)
27	Agriculture / forest transition: Atco, Waterford
28	Primary use agriculture: Pemberton and Southampton field crops
29	Primary use agriculture: Southampton dairy farms

The Coast

30	Bay and land-oriented traditional communities: Cedar Run to New Gretna, Barnegat
31	Manahawkin mixed development
32	Mixed traditional suburban: North-Forked River, Waretown, Lanoka Harbor, Toms River
33	Barrier Islands: Long Beach Island, Island Beach
34	Bay: Barnegat Bay, Great Bay, mouth of the Mullica
35	Lower Route 9: Beesley's point to Cape May
36	Route 49: head of Tuckahoe River, Marshallville, Petersburg
37	Route 47 corridor: Dennisville, Delmont, Eldora Remnant agricultural suburban: Cape May County
38	Marshes: Atlantic coast salt marshes and adjacent wetlands
39	Delaware Bay marshes
40	Tidal navigation corridors: Great Egg Harbor, Mullica River
41	Maurice River Township core: Port Elizabeth, Bricksboro, Dorchester, Leesburg, Heislerville

residents from the policy areas serve on a subregional body which, in conjunction with a staff member, would recommend performance standards and resource programs for that subregion. The subregional plans would have to fit within the legislative framework of the Pinelands Act so that the subregions would in fact sharpen the understanding of the elements of the whole national reserve rather than destroy its integrity. The Pinelands Act does call for conformance to a regional plan by each municipality, a process that requires each town to modify its master plan and developmental ordinance. If the towns were conforming first to subregional plans, which in turn related to the whole region, they would be conforming to more responsive standards as contrasted to blanket zoning regulations. That process has the added advantage of flexibility in terms of a town's response to changing conditions and preservation of open space and resources. Conformance through subregions would avoid such problems as those now faced by Hamilton Township in Atlantic County, which must accept a density of

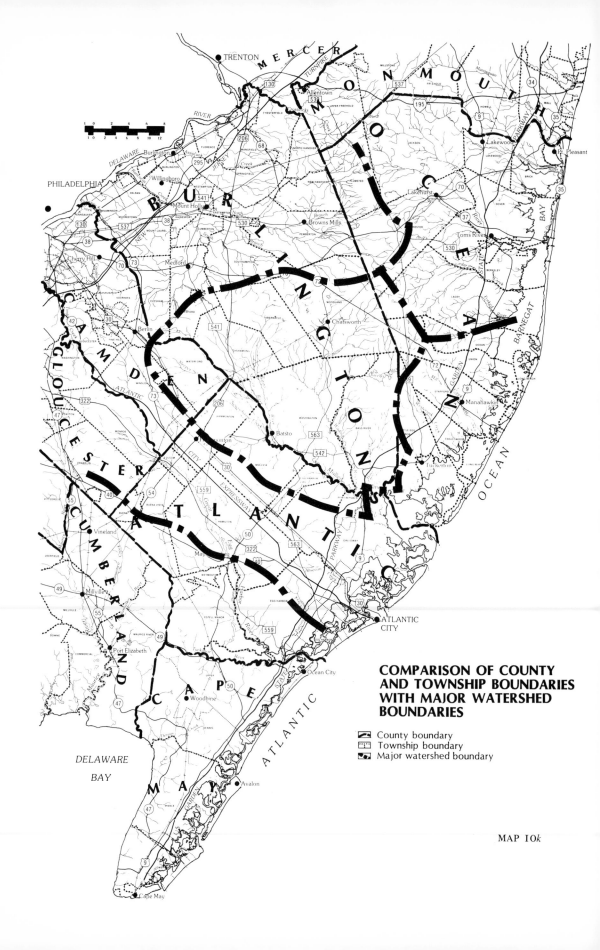

**COMPARISON OF COUNTY
AND TOWNSHIP BOUNDARIES
WITH MAJOR WATERSHED
BOUNDARIES**

County boundary
Township boundary
Major watershed boundary

MAP 10*k*

development it does not want, for which there is no present need, and which will destroy the town's character and economic structure by requiring capital improvements, which Hamilton cannot afford.

Last, use of subregions would be in the best interest of outsiders as well. If, as we contend, some of the essential character of the Pine Barrens lies in its mosaic of mixed uses, cultural features, and ecological diversity, these aspects would be well preserved for scientists and recreationists. Uniform-use zones do not engender diversity. Subregional responses to regional goals, however, would help preserve the essential character partly because they come from the local people and partly because they help avoid the artificial quality of a plan and its resulting landscape which is not based on socionatural information.

Thus our first proposition of regional control and decentralized planning utilizes the fruits of anticipatory research to identify the relevant local framework for planning. Our second proposition suggests that within these policy areas and subregions planners could offer a range of alternative preservation and development strategies but that some would be better adapted to the particular subregion. An understanding of the changing balance between family and community life and resource availability, technology, and markets can provide such insights.

PROPOSITION 2

Select when possible management strategies best adapted to the social, economic, and ecological arrangements in a subregion.

Selection of appropriate land-management strategies depends on a planner's analysis of socionatural conditions, the least known of which relate to the changing balance among family, community life, and outside pressures; resource availability; and changing markets and technology, as well as a knowledge of the range of economically feasible and legally defensible tools. Failure to understand one aspect might jeopardize the strategy one chooses.

Many local residents deeply resent intrusion into their lives. Pineys, like other groups, are alienated from the political process, feeling less and less power over their own lives and landscapes. Zoning laws may well curb the appetites of rapacious developers, but at the same time, the law cannot recognize the needs of old and established residents of the Pines.

Planners cannot hold back time; neither can many buildings be preserved or frozen as museum pieces without wasting money and becoming culturally precious. But planners can ease the transition for many people, and for others reserve a place where they can pursue their old life-styles instead of those of the postindustrial era. Since the Pine Barrens, like many other rural areas, are slowly and consistently changing—part old, part new—management strategies that fix people and place in time, by definition, will be artificial.

Local people expect change, and while those in the Pine Barrens resent a loss of local power, they even more strongly disdain the inappropriate nature of many of the management tools in the Comprehensive Management Plan. Let us take the case of Russ Clark, former mayor of Hammonton, fruit and vegetable farmer, proponent of agricultural preservation and forest protection, the original Atlantic appointee to the

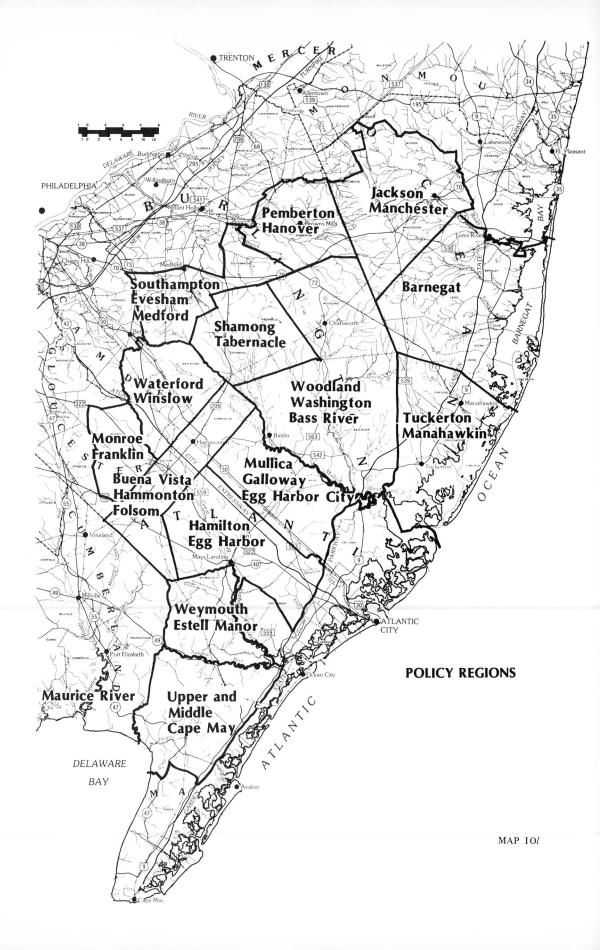

POLICY REGIONS

MAP 10*l*

Pinelands Commission, and an outspoken opponent of the CMP, who lasted on the commission only six months because his participation was costing him many of his friends. Had Russ's voice been heard during the creation of the CMP, his insights could have helped guide the planning process.

Following sound procedure, the Pinelands Commission instituted a growth-management plan to protect and enhance the unique character of the region by designating new regional growth areas to decrease pressure for development of farmland, since the regional growth areas would receive capital-investment priorities and agricultural lands would not; developers' long-term investments would, therefore, be disadvantageous in agricultural sections. Russ Clark likes this idea, since he is dedicated to farming, but he does not like the specific management techniques of the CMP. He wonders: Will it be a permanent agricultural reserve? Will there be compensation for the taking of development rights? Will regional growth areas impinge on farmlands? And if so, where?

MAP-10l

County	Historical/ Ecological/ Hydrologic Unit	Policy Region	Major Use
Atlantic	Corridor	Buena Vista/Folsom/Hammonton	Leasing patterns of crop farmers
		Egg Harbor/Hamilton	Leasing patterns of crop farmers, blueberry
	Corridor (central)	Egg Harbor City/Galloway/Mullica	Leasing patterns of crop farmers, blueberry, maritime
	Southern Pine Barrens	Estell Manor/Weymouth	Blueberry, maritime
Cape May	Southern Pine Barrens	Upper and Middle Cape May	Forest, agriculture, recreation
Cumberland	Southern Pine Barrens	Maurice River	Maritime, forest
Gloucester	Corridor	Franklin/Monroe	Leasing patterns of crop farmers
Ocean	East Coast	Jackson/Manchester	Watershed of bay
		Barnegat	Watershed on bay
		Manahawkin/Tuckerton	Watershed of bay
Camden	Corridor (central)	Waterford/Winslow	Leasing patterns of crop farmers
Burlington	Central Pine Barrens	Bass River/Washington/Woodland	Cranberry
	Shatterbelt	Shamong/Tabernacle	Leasing patterns of crop farmers
		Hanover/Pemberton	Leasing patterns of crop farmers
		Evesham/Medford/Southampton	Leasing patterns of crop farmers, development

NOTE: County boundaries are maintained due to county representation on regional commission, and within the county group townships by contiguous land and water use patterns, political economy, and watersheds.

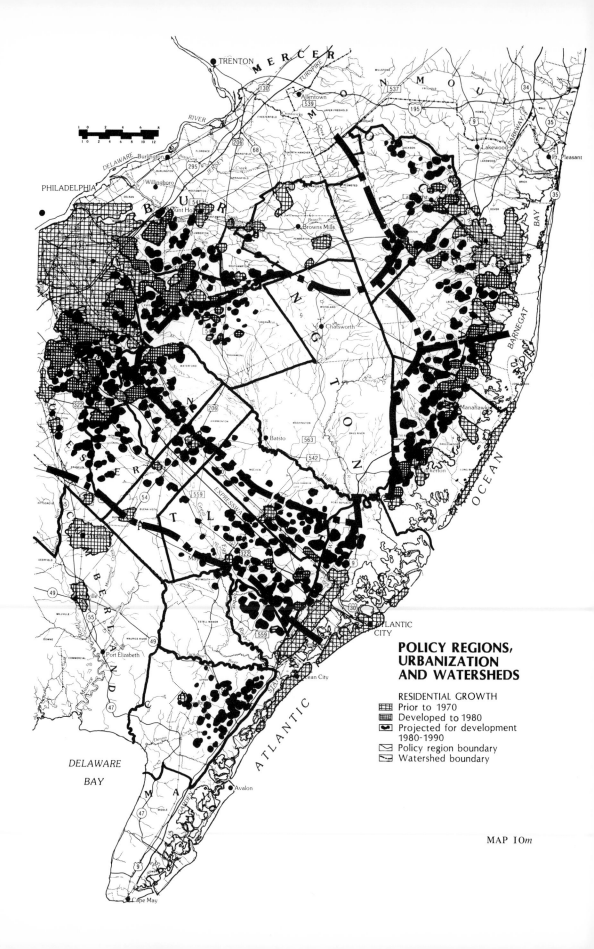

**POLICY REGIONS,
URBANIZATION
AND WATERSHEDS**

RESIDENTIAL GROWTH
Prior to 1970
Developed to 1980
Projected for development
1980-1990
Policy region boundary
Watershed boundary

MAP 10*m*

In answer the Pinelands Commission (1980, 210) decided that farmers in Agricultural Production Areas would receive two development credits for "each thirty-nine acres or the appropriate fraction thereof."

The Pinelands Development Credit Program is supplemental to the regulatory elements of the plan and provides an alternative use to property owners in the Preservation Area District, Special Agricultural Production Areas, and the Agricultural Production Areas. The program allocates to land owners in these restricted areas credits which can be purchased by land owners in growth areas and used to gain bonus residential densities. The credits thus provide a mechanism for land owners in the former areas to participate in any increase in development values which is realized in the growth areas.

From Clark's view there are problems. First, the system is a substitute for the direct sale of land. Clark, under the system, is therefore looking not for a willing buyer of land, but for someone to buy a credit. If the economy does not produce a need for greater densities in growth areas, he may not be able to sell his credit or, therefore, raise capital to meet expenditures for new machinery if he cannot sell off some land. Second, Clark knows that all his land is not good for fruit and vegetables, but all of it comes under the classification of Agricultural Production Area. He has his "sugar sands," and he cannot make money on them. Third, he is worried about financial security for his family. In his view, the system is too rigid because it does not account for the problems of the farmer or the capabilities of his land.

Farmers are not the only residents worried about Pinelands development credits, and the marketability of the credits is only one problematic issue. Rubinstein (1983, 282–83) suggests three other problems:

Second, the number of communities willing to receive these credits, and increase their overall density remains to be determined. They can only be applied within the Pinelands, and only where their introduction exceeds neither local nor Pinelands density requirements.

Third, the market value of a credit has not been determined as yet, and there are too few transactions to predict its level. A bank proposed for the purchase of credits from owners who would suffer a hardship by the restriction of their land as yet remains in the legislative process, its outcome undecided. [A bill to establish a Pinelands Development Credit Bank was vetoed by Governor Kean in January, 1984.] While the Pinelands Commission originally estimated their value at 20,000 to 30,000 dollars each, Burlington County has recently bought seven credits at 10,000 dollars each in three separate transactions. On that basis, farmers have been outraged to note that such a value represents four hundred dollars per acre for arable land—a value they say is far too low to represent compensation for the loss of development rights on their land—a value used as collateral at the beginning of each season, when they must finance equipment, supplies, and seed, pending receipt of the value of their crops.

A fourth issue concerns the stripping of the development potential of the land which may significantly alter its resale value or marketability to new owners should the owner of record choose to sell it. There is as yet no information concerning the secondary transaction and its value.

Planners, sensitive to the natural patterns of the landscape and the impact of development credits, might have alternative proposals. Indeed there are numerous alternatives. Coughlin and Keene (1981) present a compendium and evaluation in their National Agricultural Lands Study (table 4).

Perhaps the system best adapted to the Hammonton region, with its stable family farms, mixed land capabilities, and adjacent suburban pressure, would be a form of agricultural zoning that allowed nonfarm development only under certain conditions. In the vocabulary of the Hammonton farmer, the conditions are as follows: (1) development can occur on soils unsuited for fruit and vegetables; (2) development cannot interfere with agricultural activities; (3) development can occur on plots too small to farm. To translate these ideas to a zoning map, one would map sugar sands from color imagery and field checks, areas not subject to prevailing winds during spraying season, and other land cut off by roads and right-of-way that is too small for commercial agriculture. Almost every farm operation in Hammonton has these attributes. After such areas are identified, the commission could set minimum lot sizes and require compliance with all Pinelands standards. Future development would be limited only to mapped areas, which would, in effect, contain the development value of the farm. Once sold for houses, no further development would be permitted. Of course the remaining productive land could be sold at an agricultural value.

What are the benefits of this proposal? Since the Hammonton farmer's land sales are not contingent on activity in other parts of the region, he retains control of the sales. In the short run, he may receive more for a lot and house than for a development credit, but in the long run development credits might be worth more. Farmers would have to choose. Limited but available acreage for new development means a farmer has an option to provide security for his family. A frequent complaint from Hammonton women is: "What will I do if my husband dies?" The nest egg created by this process provides security but does not destroy family options for surrounding families because it does not raise land values or provide for suburban intrusions. This type of "conditional zoning" will probably work well in an area of stable family farms where farmers continue to purchase land for agriculture because there is no stampede to sell large plots to developers. In Hammonton this mechanism will preserve agricultural resources, but allow for the flexibility needed to continue farming. In other crop areas, such as Southampton Township, where the ownership pattern has shifted toward speculative ownership, and the remaining farmers lease large acreage, perhaps a more stringent, rigid program, like development credits, would be useful, especially if the transfer of those credits were done on a subregional basis. The difference between speculative and farming zones (Southampton and Hammonton) would be defined by historic considerations: the percentage of land owned by residents, the difference between the value of farmland with and without development, the amount of development pressure on the zone, and the relative importance of farming to the economic life of the local area.

This proposal builds on the strength of a stable, family-farm subregion and allows for some land conversion because the farmers have traditionally sold off small plots to raise capital to continue long-term investment in agricultural lands. This is the intent of Pinelands legislation, and it combines the farmer's pragmatic view with the prescriptive model of the planner. It relies on site-specific understanding of the physical landscape and the options of farmers.

TABLE 4. *National Agricultural Lands Study Options*

General Purpose of Program	Type of Program		
	Land Use Programs	Tax Incentive Programs
Reduce Relative Attractiveness of Farming Area for Development	Comprehensive Planning and Facility Location	Capital Gains Tax
Offset Additional Burdens on Farmer Caused by Urbanization	Agricultural Districting	Preferential Assessment for Property Tax
	Prohibition of Local Nuisance Ordinances	Deferred Taxation for Property Tax
	Right to Farm Legislation	Inheritance and Estate Tax Benefits
		Income Tax Credits
Prevent Changes in Land Use	Restrictive Agreements	
	Agricultural Zoning	
	Development Permit Systems	
	Purchase of Development Rights	
	Purchase and Resale with Restrictions	
	Transfer of Development Rights	

SOURCE: *Coughlin and Keene (1981), 39.*

Our first two propositions provide examples of the contribution of the combined user/planner perspective for the organization of a planning process and the implementation of conservation goals. The next three propositions deal with site selection, the physical and biological processes of land management, and the question of aesthetics. Our first two propositions may be of more interest to traditional policy planners, while our last three are important to physical planners and landscape designers. Again the propositions augment standard planning and design processes (Koh 1982).

PROPOSITION 3

Recognize landscape patterns of use and tradition as a basis for siting new uses.

The Comprehensive Management Plan does seem to recognize these considerations.

The Commission's critical areas study was completed by the firm of Rogers, Golden, and Halpern. The objective was to develop and execute a method for establishing a ranked list of critical areas in the Pinelands. The first step was the definition of significant natural and cultural resources. Significant resources are those which are identified as being necessary to maintain the essential character and integrity of the existing Pinelands environment. They are recognized as being valuable to the public in terms of economic, public health, safety, recreation, aesthetics, research, or education. Natural resources are the biotic element of air, water, soil and the biotic elements of individuals, species, populations, communities, and ecosystems. *Cultural resources consist of archeological or historic sites of national, state, and local importance as well as sites which are of value to a local community's way of life*. (Pinelands Commission 1980, 183; italics added)

Although this is a laudatory statement, unfortunately no definitions can be found in the CMP for sites that are of value to a local community's way of life, and a critical-areas study should rely on cultural as well as biological resources. That kind of information is available, and perhaps the reader has already formed a picture of such cultural resources from the spokespeople in *Water, Earth, and Fire*. The Manahawkin-Tuckerton subregion can serve as a good example of cultural-resource documentation for critical-areas assessment because it includes most of the resource uses found in other Pinelands subregions. As noted in previous chapters, we went into the field with area residents to observe and participate in their activities. We listened to their stories and songs, interviewed resource managers, and read the published literature. What follows, then, is our translation of community life expressed through the seasonal economy and collective memory onto maps and into policy statements.

Merce Ridgway, bayman, carpenter, and folksinger, repeats an old saying heard along the coast: "You can always make a buck on the bay and a dollar in the woods." Like the following song, it is indicative of the traditions of the place.

"My Jersey Home," Original Version

© 1980 by Merce Ridgway, Jr.

1.   I've got a song to sing
     There's something I would say

And it moves my heart
As I stand here today
Chorus: Folks I sure do miss My Jersey Home
             I miss the oaks and pines
             The meadows I once roamed
             It's like I moved away
             When I remember yesterday
             Folks I sure do miss My Jersey Home

2.   Where farmer Gray
     Once grew delicious corn
     The sawmill where my first work boat was born
     There are houses there today
     And the deer are chased away
     Folks I really miss My Jersey Home

Repeat chorus:

3.   Whoever thought
     They'd want our doggone pines
     With its sandy soil
     Mosquitos gnats and flies
     They even filled in parts of the bay
     To build houses
     Where rich folks play
     Folks I sure do miss My Jersey Home

Repeat chorus:

4.   I remember fishing on the bay
     Clams and oysters, scallops we enjoyed
     But folks I have to say
     They are mostly gone today
     Folks I really miss My Jersey Home

Repeat chorus:

5.   I'm not trying to be funny
     Or be smart
     And I'm singing from
     The bottom of my heart
     In hopes you'll understand
     How much I love this land
     Folks I really love My Jersey Home

Traditions that help stabilize economic and community life along the coast are: woodcraft, clean water, agriculture, management of fire, fisheries, wildlife, and forestry resources, free access to diverse environments, and a slow rate of growth and change. An understanding of these traditions and their spatial patterns leads to an understanding of social and physical health.

Social health is derived from an adequate source of income, a safe and secure environment for home, business, and recreation, and recognized and constructive patterns of leadership, all of which contribute to a community's ability to respond to stress and solve problems. Physical health comes from management of the environment in order to enhance the life-giving processes that provide safe, secure, and productive environments for all human uses. The behavior of insiders most closely approaches, although it does not guarantee, the ideal of physical and social health. Certainly we are talking about an ideal, and, when only a minority work toward a healthy environment, the existence of compatible use patterns is threatened from both within and without. Both insiders and outsiders need to understand that heavy human use can occur without environmental and social degradation, as long as it responds to the processes that create healthy conditions. Not everyone will get what he or she wants; trade-offs must occur, but we still need an ideal to reach toward.

As a practical planning tool we must know the relationships between the traditions of compatible use and the natural and built environments of the Barrens. Critical factors may occur together in space or be linked by natural or cultural processes. When mapped, these factors provide the structure of natural and social health. This process of landscape interpretation is important, because it synthesizes the users' view of place with the scientists' operational model of the environment. The layer cake of the Barrens can be interpreted for this structure of stability; it is the art of the ecological planner trained in ecology, the reading of landscapes, history, and anthropology.

This art does not always translate into perfect results because compromise between the viewpoints of the users and the scientists does not always occur. Two short examples can illustrate the problem. First, in the Pines, as in other parts of the country, hunters often refuse to shoot female deer because of their bias against killing females of any species, even though wildlife biologists urge doe seasons to check the over-population of deer. Second, an intriguing study by James Acheson and Robert Reidman (1982) describes suggestions by marine biologists for increasing the limit of the size of Maine lobsters taken by lobstermen because too many small lobsters are taken before they become large enough to breed. The results of the study graphically highlight problems that resource managers often confront; they are particularly relevant to the problems of commercial clammers in the national reserve, who are faced with the same kind of dwindling marine resources as are Maine's lobstermen.

According to our best estimate, fishermen will get a 13% return on the financial sacrifice they make during the 5 years of change in the legal measure [of lobster size]. Return on investment in growing industries in the United States was only 10–12% during 1979, and this is considered good. Given these figures, fishermen probably would be rational to support an increase in the carapace [upper body shell] measure. Two important caveats should be added. First, there is no guarantee they would receive 13%; some of our data are uncertain enough that smaller or larger returns might materialize. Second, the inevitable short-term sacrifice will be borne by fishermen now in the business, some of whom may not be around to reap the long-term benefits; the benefits will accrue, in part to newcomers. Thus, to individual lobstermen, the financial merits of a change in minimum legal carapace length may appear dubious. From the point of view of the state of Maine as a whole, one can make a stronger case for increasing the measure. Our data indicate that an increase in the measure likely will bring some increase in revenues in the future. We are not certain which fishermen will gain the benefits but we know that fishermen in aggregate will gain financially over the long run. (10–11)

Analysis of Manahawkin-Tuckerton Subregion. While traditional uses and sustained resource yield are not always compatible, we must, nonetheless, examine the landscape factors most closely associated with such traditions. The factors mediating the struggle between cultural point and counterpoint are the planner's best hope of achieving balances and understanding the substance of conflicts which we have referred to so often in this book.

What follows is a series of maps of the Manahawkin-Tuckerton subregion that locates critical uses and landscape factors. The maps include data on woodcraft (map 11a), shellfishing (map 11b), agriculture (map 11c), access points and corridors (map 11d), fire fighting (map 11e), and clean water (map 11f).

Woodcraft relies on the continued availability of and accessibility to woodlands, and, therefore, depends on forest management, sand and corduroy roads, beach access for driftwood, and traditional village patterns. The tourist trade and boatbuilding markets support woodcraft, as do furniture buyers and duck hunters in search of decoys. Woodcraft is also dependent on fire fighting, and there are cybernetic relationships between forest management and the mosaic patterns of the forests. Fire history, wet to dry edges, forest succession, and placement of roads, villages, dams, canals, and dikes are essential to fire fighting and forest management.

Crop agriculture in this subregion is confined to the less well-drained soils of the Cape May and Beacon Hill formations and to intensively worked sandy plots with additions of organic material. The resource is small, intensive, and scattered in association with villages, coast roads, and the remnants of a once-thriving chicken industry. Some small blueberry patches are found along with several small cranberry bogs, under constant threat of urbanization. Even small agricultural fields are so vital to the subregion that some residents suggested that forest land be cleared for new housing before the fields are used. All produce is consumed locally, much of it going to vacationing visitors.

Barnegat, Manahawkin, and Little Egg Harbor bays are prime hard- and soft-clam areas, and shellfish beds depend on the protection of both the bays and adjacent upland watersheds. Hard clams thrive on almost any bottom material at thirty-foot depths in water where salt concentrations are more than fifteen parts per thousand. Soft clams (also called steamers, pissers, or Ipswich clams) can tolerate salinities as low as five parts per thousand and are generally found in water less than three feet in depth. Commercial fishermen rely on a system of lease lots, mostly located in protected areas near shorelines. Clammers nurture their lots, and some have even begun their own seed-clam programs that require year-round attention in order to raise clams from their earliest stages. Clams are presently declining because of both the overuse of commercial operators and the hordes of recreational clammers. Important landscape elements for the clam industry include docking areas and road access to the shore. What is needed in this area is a study similar to Acheson's and Reidman's (1982) to determine the parameters of the problem and alternative solutions.

Access to upland areas is important to both insiders and outsiders. The relatively clear demarcations between vegetation types on the wet-to-dry continuum provide extensive natural pathways, which, combined with the road network, allow people and animals entry into all Pinelands habitats. There are, in addition, fire lanes, cranberry dikes, and abandoned blueberry fields, all of which provide access for recreation opportunities.

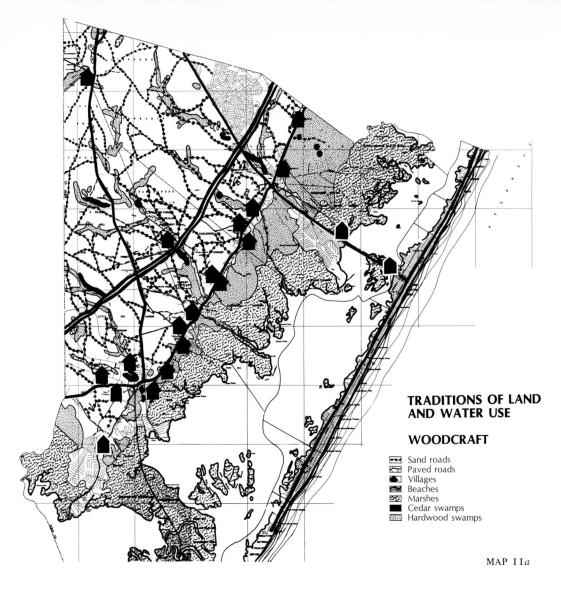

TRADITIONS OF LAND AND WATER USE

WOODCRAFT

⊟	Sand roads
⊞	Paved roads
◼	Villages
▧	Beaches
▨	Marshes
◼	Cedar swamps
▒	Hardwood swamps

MAP I I*a*

All land and resource uses in the Pines depend on clean water. In the past, everyone used the soil and marshes to purify wastes. Today sewerage systems are necessary, and land-use plans must ensure adequate recharge to aquifers by keeping large areas undeveloped; of particular importance are the well-drained soils, flood-prone areas, streams, lakes, bogs, areas of interbasin transfer, and headwater regions.

These patterns reveal the cultural and natural structure of equilibrium and provide a basis for a human ecological response. The pattern is complex but not prohibitive; it can help choose the best sites for new development. Let us first examine appropriate sites for new commercial and residential use, and then develop management practices that respond to the patterns.

Responsive Siting of New Suburban and Urban Development. New urban development is both a blessing and a curse to coastal sections of the Pines. If placed with respect to socionatural processes, and if the rate is slow and controlled, it can be a great benefit. If new development destroys compatible land-use patterns and overwhelms traditional society, no one will benefit. Siting for new urban and suburban uses must

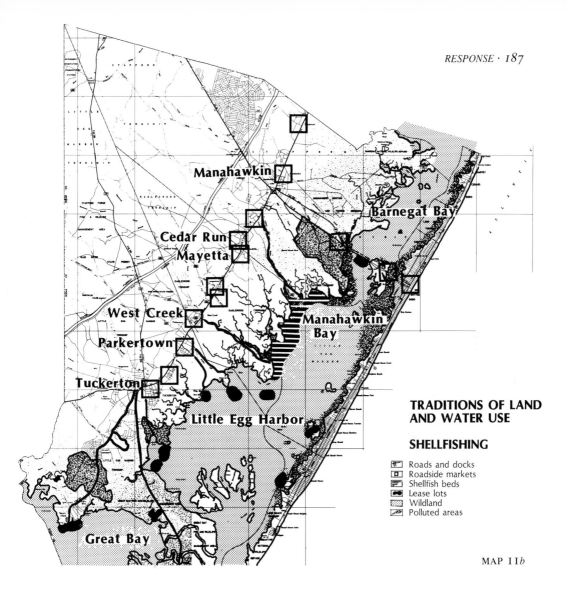

**TRADITIONS OF LAND
AND WATER USE**

SHELLFISHING

- Roads and docks
- Roadside markets
- Shellfish beds
- Lease lots
- Wildland
- Polluted areas

MAP I I*b*

maintain water quality and access to resources; it must be placed in those areas least susceptible to fire and must not overburden the services of the older community. The patterns of landscape equilibrium provide many suitable sites.

The coastal townships' boundary lines have historically followed the watersheds, and each town has its own stream pattern. The farther down in the basin new urban uses occur, the less deleterious the impact will be on the surface water system. In any coastal watershed, first-order stream branches are the base level for the system of wet and dry edges and corresponding set of human uses. Thus, if new urban uses occur below the last of the first-order tributaries, these uses will occupy sites that have the least disruptive effect on the system of resources and access to them. If at all possible, headwater areas must be left untouched so they can remain linked and continue to serve as the supportive upland for the coastal strip and bays. The concept of downstream siting is the first principle derived from patterns of equilibrium.

Second, all downstream locations are not equal. Both sides of a stream should not be utilized, for use cuts off human and animal access down the corridor to the wooded swamp or marsh. Therefore, each stream mouth should be used in an alternating

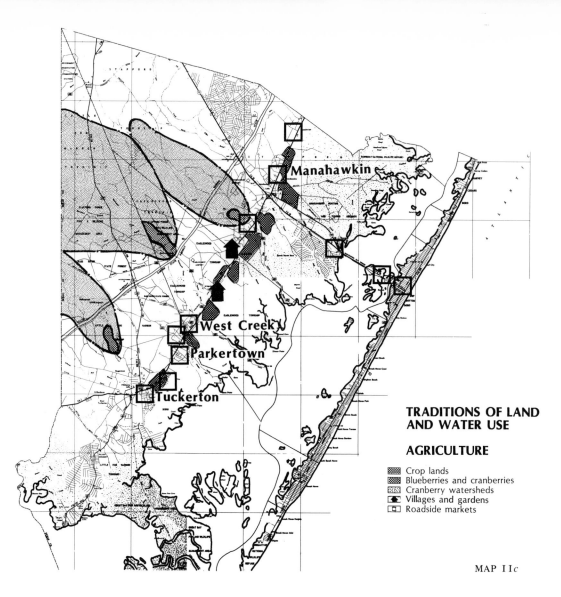

TRADITIONS OF LAND
AND WATER USE

AGRICULTURE

▨ Crop lands
▨ Blueberries and cranberries
▨ Cranberry watersheds
◨ Villages and gardens
▢ Roadside markets

MAP IIc

fashion, which results in the opposite side being left open each time development is sited near a riverbank. Siting near, but not on, the creek is also critical. Even if development occurs on a downstream site, it should respect the wet-to-dry continuum of the shallow valley. Thus the limit of built structures should be the first sand road encountered on the dry upland. Less intensive uses should grade off toward the wetland corridors. Firebreaks, fields, or gardens should be the new uses nearest to the stream.

Again, this is an ideal because land-ownership patterns, social pressures, and political considerations can undo the best intentions or the most compatible uses. However, these selected locations have considerable benefit for new uses. First, the downstream position is in a less fire-prone area than the upland. The water table is somewhat close to the surface, and to the east are the marsh and bay, which are not fire prone. The downstream position also maximizes fire protection from the swamps found on all the untouched, first-order tributaries of the upper basin, all of which lie west of the new development, and it is from the west that spring fires descend. Further, the exact location of the new development can take site-specific advantage of the juxta-position of the nearest tributaries and the main stem of the basin. It is quite possible to

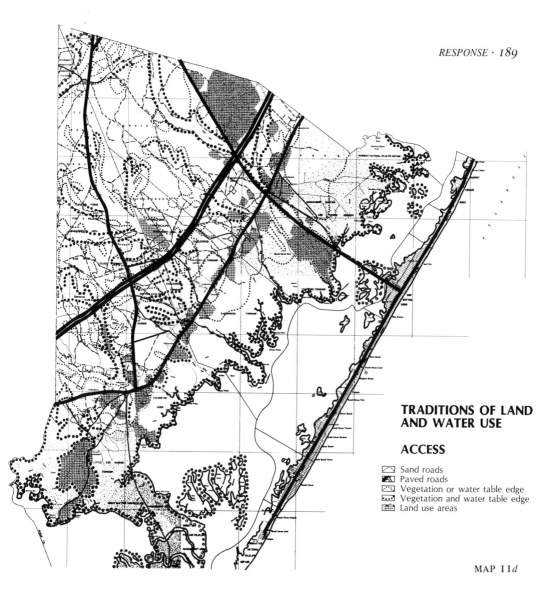

**TRADITIONS OF LAND
AND WATER USE**

ACCESS

 Sand roads
 Paved roads
 Vegetation or water table edge
 Vegetation and water table edge
 Land use areas

MAP IId

site new uses on dry uplands that have firebreak swamps on two sides. Fire protection is only one of several benefits; conducive climate and cost savings for water recharge are others.

Because there are almost two hundred frost-free days per year on or near the coastal strip, less energy is required to heat homes than in the interior Pine Barrens. Offshore breezes also provide a measure of summer comfort, which means a lighter cooling load; less energy use is of particular benefit to new developments.

As the micro climates save money, so do the well-drained soils of the sites. All new development will have to recharge storm-water runoff on site, which has the advantage of easy on-site recharge without the detriment of too rapid a flow through droughty upland soils. New development will cause more runoff because many of the best locations are forested, and tree clearing will change the balance between infiltration and overland flow. Clearing of trees creates firebreaks and relieves development pressure from the small fields of traditional villages. These are beneficial trade-offs.

New sites for new developments will not necessarily conflict socially or aesthetically with the older established settlements. The Pinelands plan requires infill of existing

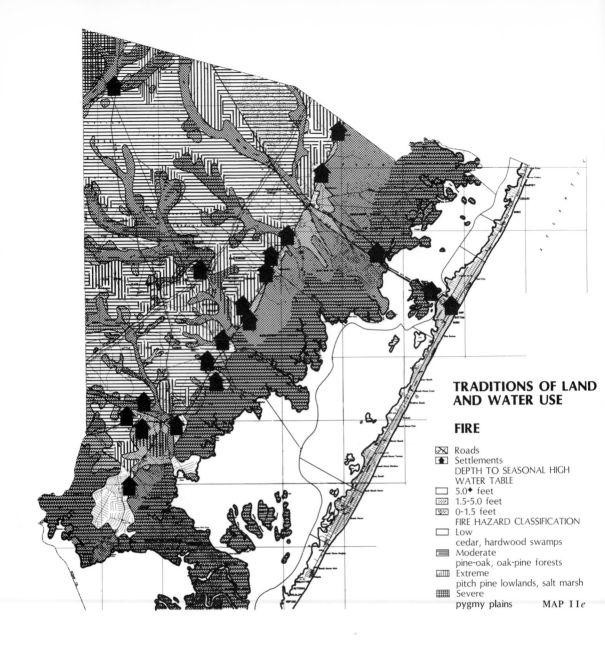

**TRADITIONS OF LAND
AND WATER USE**

FIRE

⊠ Roads
⬛ Settlements
DEPTH TO SEASONAL HIGH
WATER TABLE
☐ 5.0+ feet
▨ 1.5-5.0 feet
▨ 0-1.5 feet
FIRE HAZARD CLASSIFICATION
☐ Low
   cedar, hardwood swamps
▤ Moderate
   pine-oak, oak-pine forests
▥ Extreme
   pitch pine lowlands, salt marsh
▦ Severe
   pygmy plains   MAP 11*e*

settlements, but this may not prove beneficial. Instead, we recommend evaluation of sites based on patterns of equilibria, which would be different for each subregion. At the same time, construction work will provide seasonal labor for many of the older residents and investment for the local banking and merchant community. Newcomers will not, perhaps, be greeted openly. It will be helpful for the new residents to have their own clusters of houses where they can start their own associations of homeowners and contribute in a block to the local rescue and fire-fighting companies. New homeowners may very well want to start their own fire companies. Since some of the downstream sites are a square mile or more, a new cluster of houses may also provide the opportunity for lake construction. If these areas correspond with old sand and gravel quarries, as some do, then land reclamation can create an attractive site for new settlement.

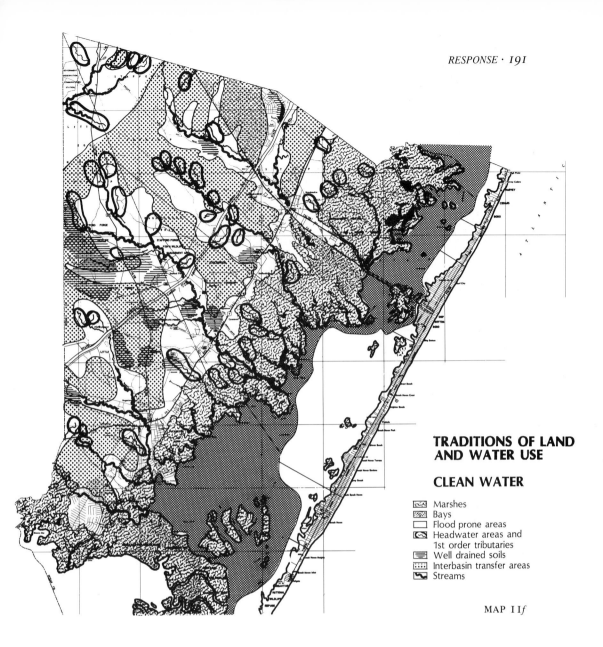

TRADITIONS OF LAND AND WATER USE

CLEAN WATER

Marshes
Bays
Flood prone areas
Headwater areas and
1st order tributaries
Well drained soils
Interbasin transfer areas
Streams

MAP 11*f*

A lake provides many benefits. First, residents will have to organize to care for it; dams and beaches need maintenance, and seasonal water quality and lake levels require attention. Second, the lake can serve as a means of integration between the older and new residents. If the new residents are informed of the need, they can provide recreation for local children, which will endear them to the parents and grandparents of older residents. Lakes have usually been seen as symbols of exclusivity, but if they are used to help integrate a community, they will contribute to a more healthy social environment. Location in the downstream position can utilize the opportunity of a water table relatively near the surface for lake construction.

Map 11g shows available downstream positions and their suitability for new development. The best areas have a concurrence of fire protection, downstream position, the possibility of abandoned quarries, and access to local and regional roads.

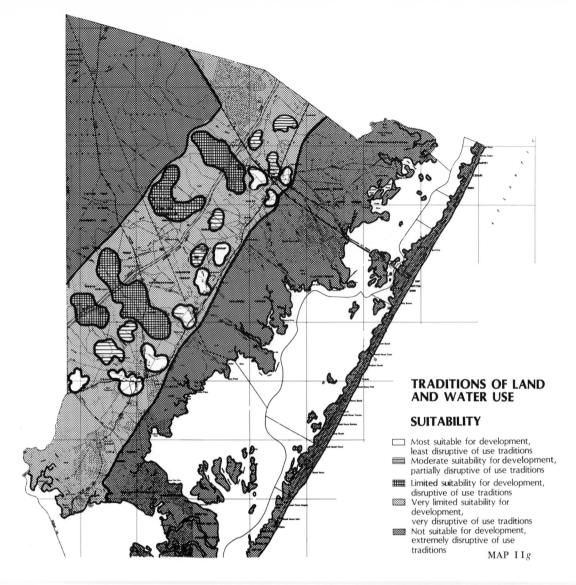

**TRADITIONS OF LAND
AND WATER USE**

SUITABILITY

☐ Most suitable for development,
least disruptive of use traditions
☰ Moderate suitability for development,
partially disruptive of use traditions
▦ Limited suitability for development,
disruptive of use traditions
▨ Very limited suitability for
development,
very disruptive of use traditions
▩ Not suitable for development,
extremely disruptive of use
traditions

MAP IIg

Less suitable areas have only the downstream position and fewer amenities. Some areas are indicated in more upland areas that do not violate access, but are further up in the basin than ideal siting would provide. All townships have at least one area larger than one square mile that meets the necessary criterion of equilibrium.

These locations are a guide to the use of growth-management tactics to insure a slow rate of growth that will not overwhelm the local community. The townships and county can use their capital budgets to stage the extension of services to these areas and regulate the pace and location of development. Preferred sites receive services first. Scattered rural development outside these areas should also respond to access, water quality, and fire protection.

Unfortunately, standards for new developments in the Pinelands plan read like those in a zoning ordinance for a million-acre site. The standards are so strict, and density requirements so rigid, that it is almost impossible for a town, even in the conformance process, to create an imaginative plan that responds to socionatural processes. Particularly onerous are development standards for the preservation zone, which require the developer to have 3.2 acres and to "demonstrate a cultural, social or economic link to the essential character of the Pinelands'' by being a member of a two-generation extended family that has resided in the Pinelands for at least twenty years or to show economic dependence on a resource-related activity in the Pine Barrens (Pinelands Commission 1980, sec. 5-302). Our discussion of point and counterpoint shows that membership in a two-generation family is a patently ridiculous, irrelevant, and discriminatory test for the preservation of Pinelands landscapes. The question is where uses should occur, not who should participate. The planners should have sufficient knowledge of the socionatural patterns to locate the places most important to these people.

It makes more sense to tailor housing and development standards to subregions. Management techniques that do not fit the landscape will either be rejected like foreign matter from a body, or imposed to few people's benefit. Not even outsiders are helped by the so-called Piney exemption, because it does not foster an indigenous or aesthetically pleasing landscape. It may prevent major developments, but not scattered single dwellings, which are, in any case, part of the historic pattern of the Pines.

Let us now turn to ways planners can weave tradition, local skills, and actors into the changing landscape. Many traditions are indicative of the collective memory that provides highly developed skills to help manage resources.

PROPOSITION 4

Incorporate local skills and actors into site management and mix these with applied science.

The CMP contains hundreds of standards and regulations for the use and management of the Pinelands. Few, if any, are derived from an understanding of the seasonal cycle and the collective memory. Throughout the rural regions of America land users have developed skills adapted to the ecological and economic realities of the region, and these are useful to the conservation and future development of rural America.

Fire and Forestry. The emphasis in the CMP with respect to forestry and fire is on the protection of water quality and property. Forestry regulations cover erosion, sedimentation, and revegetation. A classification of fire-hazard areas relates to the width of the necessary firebreak needed to protect human habitation. Yet an imaginative program of forestry and fire management could, in fact, enhance clusters of human use, protect the environment, and guard property.

Fire management is crucial to the siting of new development and the maintenance and enhancement of aesthetic, recreational, and historic resources. Of all the facets of Pine Barrens life, adaptations to fire are, perhaps, the most suitable focus for regional-management systems. Silas Little, now retired from the U.S. Forest Service and a student of Pinelands forests for three decades, believes that modern fire management, like the commercial blueberry, began in the Pine Barrens.

No resource use in the Pines occurs without regard to fire. Human settlements are sited east of the river swamps, which are natural firebreaks. Fields and gardens are man-made firebreaks. Homes are scattered and houses are constructed for easy access to them and to the roof for emergency fire fighting. Residents and state agencies protect their homes, fields, packinghouses, and hunting lodges with a three-year rotation of ground fires, called prescribed burning. In late winter, on still days, fire wardens and volunteer fire companies use creeping ground fires to burn up litter and reduce the threat of wildfire during spring season. This burn favors pines over oaks and produces a parklike forest with few understory trees and shrubs. It impoverishes the wildlife habitat for some game animals, but favors pulp-wood production.

Those who favor wildlife management advocate a five- to seven-year rotation of hotter fire that opens up the forest canopy, favoring scrub oak, grasses, and forbs. Some hunting clubs with extensive holdings use this kind of fire management.

The pygmy forests of the Plains burn on an average of every seven years without fire suppression. Some favor a three- to five-year rotation for the Plains to maximize the rare and endangered species and reduce the possibility of a Plains fire spreading to adjacent areas. Others favor wildfires that burn at will.

It is possible to create fire zoning that enhances clusters of human use. One policy need not cover all fire-prone areas. Human settlement and the headlands of bogs, blueberry fields, and agricultural buildings need a three-year rotation. Areas owned by hunting clubs, state game-management areas, and selected sections of state forests would benefit from a five- to seven-year hotter burn. Plains areas would require their own regimen. Large tracts of public and private property, suitable for pine plantations, would have three- to five-year rotations. This system opts for a diversity of interests, yet maintains the integrity of the fire-prone ecosystem and patchwork mosaic of the forest.

One need not impose fire policy from the outside because skilled local people can successfully implement it. These policies are not the work of environmentalists or bureaucrats, but long-standing local strategies. Members of gunning clubs, volunteer firemen, local fire wardens, and loggers provide a skilled local cadre to manage fires; many will volunteer if it is clear they will benefit directly from their work. Fire-management teams could receive donations from local landowners, or they could sign fire-management contracts for a fee and sell the standing burned wood for fuel.

If notices of fire-management programs could be sent with tax bills to local and absentee landowners, there is every reason to believe that area residents would be responsive to one of their most vital concerns. In their townships or fire districts, local residents could meet to decide what types of fire-management activities they wished to encourage. Government officials, landowners, gun-club members, and civic and volunteer associations could work out local fire zoning for every township.

Fire and timber cutting helps maintain the forest's patchwork mosaic, which reduces fire hazard, increases recreational and economic opportunities, and enhances community unity. White cedar plays a special part in forest management because of its ecological, economic, and aesthetic value. Given sixty years and minimal care, it regenerates quickly and should be clear-cut. But, given greed and lack of forest management, only 50,000 acres of cedar are now left in the Pine Barrens (Pierson 1979). Cedar lumber mills tend to be small, family operations, and the scarcity of the resource tends to keep them that way. Jack Cervetto is lucky to own his own swamp because he can custom-cut lumber or sell limited cutting rights to neighboring mill owners. Herschel Abbott of Manahawkin operates a mill and buys cedar from Jack. Herschel counts himself lucky to be able to rely on Jack because he generally has to buy cutting rights from other swamps to get many of his logs. Herschel just shakes his head when he thinks of the productive swamps he cannot cut because a speculator owns the land, but he knows that since many speculators do not have clear title to their lands or proper surveys, he would be partner to timber thievery if he bought from them.

Neither Herschel Abbott nor anyone else knows how much cedar is held out of cutting or management, but this situation is appropriate for government intervention. Freshwater swamps are excellent recommendations for scarce acquisition funds because such areas protect water quality and harbor habitats for endangered species. Furthermore, they are small and can produce the most valuable timber resources in the region. Hundreds of small purchases would protect and enhance a precious physical and cultural resource for the Pine Barrens, and once bought, swamps could be leased back to mill owners in a manner similar to clam lots; they could be sold on the open market with performance bonds for proper management and survey, or the state could own the swamps and establish bidding schedules like those on other state lands. A lumbering cooperative organized like Ocean Spray or TRU BLU could be encouraged so that cutting and marketing controls could be more stable than they are presently. This would also reduce the risk of timber thieving and the kind of management that presently occurs due to the negligence or purposeful mismanagement of one or two large-scale operators.

Marine Resources.   Some attributes of shellfish exploitation are comparable to those of the cedar and crops grown in the Manahawkin-Tuckertown subregion. The most obvious is the small size of the operations because the most dependable catches come from small, intensively maintained lease lots. In addition, the resource is seasonal and requires a husbandman's hand. There are few commercial clammers compared to the recreationists—only about 1,200 of the 30,420 licensed clammers in 1980—but the former group harvests by far the largest proportion of shellfish. In the subregion on an average day in 1980 recreationists landed collectively less than 200 clams, while commercial clammers landed a combined total of over 1,400 clams a day (Figley and McCloy 1980). Further, harvests have increased by 17 percent since 1978. Still, some baymen will only work the bays when times are tough. Like resources of the woods and fields, shellfish are often a fallback activity when other work is unavailable. Such resources are vital to the equilibrium of the region because they provide a cushion of security for family and communities. Management and exploitation of shellfisheries must aim for long-term maintenance for insiders and outsiders.

In the past decade, however, hard-clam populations have decreased as the number of clams harvested has increased. Baymen and biologists alike attribute the drop to pollution, which forces the permanent or seasonal closure of some of the most productive waters, and to poor recruitment of clams from smaller to larger because of bad breeding seasons. Pollution problems are attributable to coastal development, and the condemned shellfish-area charts that the state publishes annually clearly indicate this pattern. All of the developed areas close to the shoreline, those most coveted because of protection from wind and rough water, have been closed as are the river mouths leading from upstream developments and dredged lagoons of nearshore developments. The lesson is clear—keep the shore free from development or guarantee that no urban runoff reaches tidal waters. This is difficult to achieve because of the well-drained soils and proximity of the water table to the surface of the land. It must be the policy of the New Jersey Department of Environmental Protection, which manages coastal resources and land uses, to work more closely with the Pinelands Commission to develop responsive policies; heretofore, communication between the two agencies has been, at best, sporadic.

If pollution abatement and artificial propagation of seed clams do not reverse population declines, the state's Fish and Game Council will inevitably lower the daily catch limits, which, in turn, will create a major political controversy. Recreational and commercial clammers will blame each other for the drop in clam harvests. State legislators will intervene on behalf of both parties; the biologists will produce numbers showing that commercial clammers take far more clams than do recreationists, and the governor will be under pressure to change the composition of the Shellfish Council. With nature's help and successful seeding operations, this controversy may be avoided, and, if the state can find a way to subsidize commercial seeding operations, a solution might be imminent. Without effort, the region may lose its balance—its collective memory, the willingness and ability of local people to adapt and innovate, and the maintenance of socionatural equilibria.

The management of fire-prone forests and shellfisheries relates directly to the visual appearance of the landscape. What becomes apparent from traveling through the Pine Barrens, participating in its land uses, and visiting its residents is that there are highly developed, yet not easily recognized, regional aesthetic norms. To the outsider, they may go entirely unrecognized. Our last proposition deals with aesthetics.

PROPOSITION 5

Recognize local and subregional aesthetic norms in site design and regional management.

Forms of aesthetic expression commonly reveal the historical relationship between people and their environment (Rose 1979), and recognition of these forms and their expression indicates awareness of local variation from national tastes and trends. Lack of recognition suggests a failure to understand biases of national popular culture that are built into land-classification systems. This is the Pinelands Commission's dilemma. The failure of the commissioners, staff, and consultants to read Pinelands landscapes and synthesize aesthetic information is indicated in the commission's efforts to manage scenic resources.

Any observer has a predisposition toward a landscape. Stilgoe (1981), for example, documents the impact of sixteenth-century European land-classification systems on American aesthetic thinking. Until 1870, most writers and painters saw beauty in the garden and the field, but present national taste reveres landscapes without people and their artifacts. This philosophy separates people from their environment in their own minds as well as in public policy. The geographer Rees (1975, 46) summed up the problem this way:

> The taste increasingly has been for wild landscapes and in North America this has been signaled by changes in terminology: wilderness areas for parks. . . . The point is of course that the taste has been for scenery, for a view and not for landscape in the original Dutch sense of the word—of the present geographical one—which designated the whole of landscape, warts and all.

The insiders' view of the Pines recognizes warts and all. Questions about favorite landscapes elicit the statement: "Oh, I like it all." Visual senses cannot be separated from memory, smell, touch, or hearing, so residents have a more sophisticated aesthetic vocabulary than the one utilized by the commission. The CMP relied on"scenic corridors," which include paved roads and navigable streams and lakes, as well as sign ordinances to express values for the entire region. The commission based its corridors on a survey done by the National Park Service, which asked audiences to view selected pairs of photographs of a range of land-use and vegetation patterns. The results showed that most respondents preferred wooded, undisturbed, water-related scenes, all of which supported the national bias for scenery, but not landscape. Predictably, the audiences were composed chiefly of students and white-collar workers. There were no woodcutters, farmers, or baymen in the survey and very few Pinelands residents.

However, if one walks over the ground with people who live in the Pines, and if they draw their own mental maps, stories begin to emerge. One discovers places where the Indians and privateers lived, and how patterns changed. Residents do not look purely at scenery; they know the locations of old artifacts and of forge ponds or mill villages. They also relate stories about their friends and families when they were hunting or ditchdigging or huckleberry picking.

There is, furthermore, the question of biases against junk (as contrasted to litter), woodpiles, ten-foot-high wine bottles, burned and cut forest land, delapidated trailers, and nonnative vegetation. Junk shops and flea markets are as common in the Pines as in other rural areas. How does one display junk for sale without a junk pile or a sign for junk? Firewood is another local commodity dumped or stacked in nonsymmetrical ways, and it is a well-accepted aesthetic norm. What about trailers? People's aging parents live in them, or sometimes young singles or couples. Many people put them next to their house without driveways, curbs, grading, or landscaping. What about large concrete signs or statues on farm markets? Regional farm markets along the White Horse Pike support a proud collection of enormous chickens, white prancing horses, oversized tomatoes and oranges, and ten-foot-high ice-cream cones.

Finally, how can the trained observer tell where old crossroads towns or houses might have been except by looking for locust trees, catalpas, and sycamores planted to bring a bit of exotic diversity into the landscape? All of these and more need to be recognized as aesthetic resources.

A data base for regional aesthetics does not yet exist. However, the American Folklife Center of the Library of Congress has just completed a study of Pine Barrens folklife including vernacular architecture and yardscapes, as well as ethnobotany and life-styles. From these data would flow recommendations for historic districts, the maximum size of Pinelands villages, local aesthetic resources, and design guidelines to incorporate the field, woodlot, and hamlet into new road construction and new community design. It would also help to use community people in subregions to organize an agenda for preservation or restoration (Stokes 1980). In the end, the Folklife Center's studies can give real substance to the evasive nonmaterial elements of Pine Barrens culture.

Chapter 6

CONCLUSION

Like a college graduation, this conclusion is as much a commencement as an ending. Along with other planners, we are beginning to put the pieces of a complex system together so we can come closer to MacKaye's all-round revelation. The novelty of our collective attempts to reformulate frameworks is symbolized in our sometimes cumbersome language—human ecological planning, socionatural systems, topophilia.

This book might, therefore, be regarded as an extended essay, an attempt to help construct a better way to understand a place that will lead ultimately to better planning in general. The propositions we derived from the people of the Pine Barrens are the beginning of a process and dialogue to test the value of using socionatural information. Our propositions do not constitute a whole planning process, but rather document an approach that does not separate people from place along the lines that the late Angus Hills (1974) suggested in his evaluation of the relationships between people and the earth. We are trying not to replace traditional planning, but to augment its legal, political, economic, and social frameworks.

We value the concepts of ecological planning that incorporate the natural world into decision making through ecological inventories, but suggest that such inventories be expanded to include the cultural patterns of a region. We do not view such tools as visual analysis or lists of historical and archaeological sites as sufficient data. We need an early identification of subregional environmental values, land-use behavior, and agendas for the future.

We do not yet know if it is too late for the Pinelands planning process to succeed, since so many geographical, psychological, and political lines have hardened over the past four years, but being part of that process has taught us much about what can and cannot work. More national reserves will probably be established, although not necessarily in the near future, and regional planning will continue to respond to new problems.

When we plan the next national reserve, we must be able to answer more adequately the following questions: Can we write legislation to fit unexpected exigencies? or can we compose a plan flexible enough to adapt to unforeseen conditions? Can we come up with better ways to predict the economic impacts of a regional plan such as the one for the Pinelands, the Adirondacks, or the California Coastal Zone? Can the value systems of local residents be incorporated into a plan even if they run counter to the legislation's goals? or does that question point up the limitations of the legislative aides who write bills? What part does ethics play in the planning process? What are the relationships among local towns, subregions, and regions to the nation and beyond? between individuals and the planning framework?

The last two questions are perhaps the most crucial because they are at the heart of what John Bennett (1983) called the "micro-macro problem"—that is, the interrelationships among the parts of complex systems (see Appendix). In our study of the Pine Barrens we concentrated on some aspects of the problem, particularly relationships among individuals, their resources and communities, the local and state planning structure, and, to a lesser extent, national economic, social, and political trends. What planners must strive for is the creation of a larger context in which such interrelationships are revealed more accurately. If, as John Forester has pointed out (1980, 1982), information is one, and sometimes the only, source of a planner's power, then that information must be as clear, ethical, and complete as possible. Misinformation, whether intentional or not, can be both disruptive and destructive. We have tried in this work to show how important it is to understand the social and ecological networks of a region.

But we need to look even further at the region in its larger context. It is indeed important to understand that the physical and psychological health of a community is tied to its local and regional environment, but as psychiatrist and epidemiologist Alexander Leighton (1983, 2) suggested in a personal communication:

> A difficulty is that the members of a local geographically defined community—no matter how well organized—are not satisfied with what they can produce for themselves from local resources. They try to get much, much more and as a result find themselves tied into vast networks of influences from the society at large of the continent, and beyond.

In this work we have seen historical and contemporary examples of the impact of such large networks on people of the Pines. More research, however, needs to be done to understand more precisely the connections between levels.

It could well be, as Silvan Tomkins has suggested (personal communication), that, if the parts of a system have all the features of the whole, as well as their own special functions, our study of any part will reveal some aspects of the workings of the whole. This is an exciting view of complex systems; although it is by no means novel to philosophers and physicists, it may well open doors in planning. This view, consistent as it is with our understanding many processes in the physical and psychological world, will force planners to focus on interrelationships rather than discrete bits of information; it will help keep us honest by refusing to allow planning to become uniform in its approaches and implementation. Such an approach may prove frustrating because complex systems are obviously difficult to comprehend, but the frustration will stem from honest efforts to grasp a larger reality, not just from the impatience planners feel when people and communities won't do what they are told or what is in their "best interest."

What we look toward in the end is better planning through a better understanding of how our systems work as things inevitably change. The landscapes through which people move in their lives are as changeable as are the lives themselves. Why, then, should landscapes be considered a backdrop? Because they are as stationary as they are in Breughel paintings, where wonderful things happen only to the people while elements of the landscape remain static? Does our need for a stable reference point make it necessary for our landscapes to perform the function of stabilizing us? The simple fact is that landscapes change as surely as we do; what we carry with us, to stabilize ourselves, are memories of those landscapes. It is the memories that become our realities, and we try hard to mold our environment to fit and conform to them. It is time now to relate ourselves, as planners and citizens, to the larger processes of change and to allow our realities to mix with those of others.

Throughout this book we have explored the themes of topophilia, tension between insiders and outsiders, cultural point and counterpoint, and changing balances among resources, technology, markets, family, and community. Let us take one last look at the manifestations of these themes in the artifacts of the landscape and hear their echo in people's conversation.

Photograph 20 shows about one acre of land in the village of Cedar Run, Stafford Township, Ocean County. It was taken from the backdoor of Herschel Abbott's sawmill looking west across the piles of cedar logs and small fields to the outbuildings and homes of the village residents. We can use it to interpret themes in the landscape.

The cedar logs represent cedar swamps inland and the clam stakes, lease lots, and boatbuilding on the coast. The outbuildings house clam tongs, shinnicock rakes, baskets, turtle and crab pots, and fire-fighting equipment as well as storage areas for the roadside vegetable and seafood markets. Piled around the outbuildings is firewood—oak cut from the drier uplands and sent to market or sold along the roadway for home-heating fuel. The garden indicates the special relationship between residents and their soil, which, in this case, has the slightly impeded drainage characteristic of the thin layer of Cape May gravel enriched with organic fertilizer worked into it over generations. Here, then, is a selection of the products from many of the wet to dry environments, many of the salt to fresh ones, and the numerous biotic communities of the Pines. These are artifacts produced by switching from one resource to another, helping to renew those resources, and passing skills down through time. There is yet more to read.

The homes, the mill, the purple marten houses are symbols of love of place; work, play, and rest are all close by. The woodworking and carving of birdhouses, the

PHOTO 20. *Herschel Abbott's Yard*

well-kept nature of the houses, the irregular but locally known work hours at the mill all tell of a life-style chosen deliberately for its pace, security, economy, and its ties to the land and waters.

As Berger finished taking the picture, Mrs. Abbott arrived and questioned him:

ABBOTT: You're not from the Pinelands Commission, are you? If you were, my husband would get awful mad.

BERGER: No, just taking some pictures of the birdhouses.

ABBOTT: Say, I know you. You're the writer.

BERGER: Yes, I visited you about two years ago, just after New Year's, at dusk. You were opening up to let some lads unload their cedar. How is your daughter?

ABBOTT: She has her good days and her bad days.

BERGER: How is your house coming along?

ABBOTT: Oh, with summer coming on, we are trying to finish the breezeway. Sorry to be so abrupt, but those environmentalists are causing all sorts of trouble. Darn people don't know this place, and now all these regulations.

BERGER: I am sorry to hear that. What seems to be the matter?

ABBOTT: Oh, just telling us what to do—but just down there people are building new homes, in the wet, and they shouldn't be allowed to do that. They should leave such areas be.

BERGER: I guess there have to be some regulations, some controls.

ABBOTT: Yes, but we know about cedar cutting and we are not doing any harm.

What we hear is the classic point and counterpoint. We hear the disappointment and anger at the speculators who violate the water table and fill wet soils, but we also hear the reluctance to accept outside control of cedar cutting and the explicit belief that those who regulate do not know what it is like to live in Cedar Run or cut cedar for a living. To us Mrs. Abbott is saying that she would welcome a planning process that included her family and neighbors, that recognized what makes the Barrens home for people, and yet protected the environment. Should she settle for less from the planning profession?

To us this picture shows what makes the Barrens a home for many people. It presents the seasonal fluctuation of water, earth, and fire. It is a backdrop to the struggle between the community conscience and the forces of economic and social change. All this and more can be read from the landscape and interpreted from the photos by those who take the time and care to do so. Care, time, predisposition, and training that lead to information and sensitivity—these are some of the insights and qualities necessary for someone who aspires to be a resource manager and public-policy planner in the Jersey Pines or anywhere else. The interpretation of the landscape is a humane process that enriches the life of the reader and of the people with whom the planner must work in the planning and design of our world.

Appendix

Appendix. Technical Notes on Environmental Planning

Two Perspectives

Most of the material on which this study is based comes from fieldwork we did as consultants to the Pinelands Commission during the planning process from fall 1979 through spring 1980. Along with the nineteen other consultants whom the Pinelands Commission hired, we were testing our assumptions about how one investigates and plans for a region. We did not come to our work with a *tabula rasa* anymore than did others involved in the whole planning process. We had two perspectives by which we approached our study—our academic biases. The first was the concept of the socionatural system, and the second, the contrast between the scientists' overview of the region (the operational model) and the users' pragmatic view of place (the cognized model).

The cultural ecologist John Bennett is the major spokesman for the socionatural systems that result from people's adaptive strategies. In his book, *The Ecological Transition* (1976), he summarizes the major theories of human ecology (table 5). He comments:

> The causational models represented in the five positions [in table 5] seem to be three: (A) is linear causation where A causes B etc.; (B) is the feedback or systemic model in which causation is seen as a process of interdependent, mutually influencing factors—but acknowledging varying strengths among these; (C) the third model is the adaptive model which sees outcome as the consequence of human decision or choice—a model that does not reject the systemic or even the

TABLE 5. *Major Theories of Human Ecology*

1. Deterministic Anthropogeography

Environmental Factors ——— Shape ———► Culture

(That is, environment preexists culture.)

2. Possibilism

Culture Selects From ——— to ———► Create a Subsistence Style
Environment   and Other Cultural Factors

(That is, "Culture" preexists the inquiry, "it" is the basis of behavior. Objectives largely descriptive. Environment entirely subject to culture. Choice or selection is the critical step here, though it is generalized as "culture.")

3. Stewardian Cultural Ecology

A Cultural or Social   can be   The Technical-Economic Apparatus,
Factor not Involved   ———————►   and Those Social Factors Involved
With Subsistence   explained by   With It, Involved in the
System   Subsistence System

(That is, "Culture" is broken up into variables with differing causal significance. "Environment" becomes whatever is defined by the technoeconomic "core" or subsistence system, but it retains some strength as a causal agent.)

4. Cultural Ecosystemicism

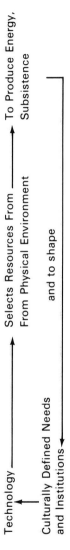

Technology ⟶ Selects Resources From ⟶ To Produce Energy,
From Physical Environment   Subsistence

Culturally Defined Needs   and to shape
and Institutions

(The components are interrelated to produce a biotic or socionatural system. No need to assign causes on theoretical grounds.)

5. Adaptive Dynamics

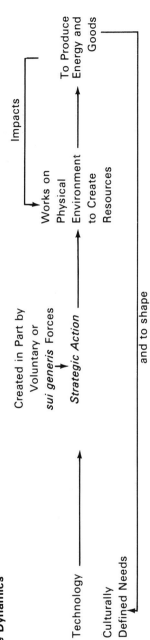

Impacts

Created in Part by   Works on
Voluntary or   Physical   To Produce
sui generis Forces   Environment ⟶ Energy and
Strategic Action ⟶ to Create   Goods
Technology ⟶   Resources

Culturally
Defined Needs

and to shape

(Similar to the preceding, but the focus of research is on the strategic behavior of the actors in making choices and modifying patterns. More feedbacks or "impacts" are acknowledged than in ecosystemicism.)

SOURCE: *Bennett (1976), 165. Reprinted with permission from Pergamon Press.*

linear, but considers these to be empirical outcomes of behavior and not inevitable workings of the phenomena involved.

We found Bennett's adaptive systemic approach the most effective for analyzing regions. The adaptive model is more conducive to analyzing "open" systems that are subject to both positive and negative feedback; such systems have no fixed response to shifts in human values, technology, environmental changes, or political or economic systems. The model attempts to integrate data on both "micro" and "macro" scales by linking the local units of production or resource use to local social organization, environmental conditions, and regional markets, and, thence, to larger systems.

However, our study of the Pine Barrens was not a complete cultural ecological analysis. We did not collect a decade of (longitudinal) observations on changes in resource use. We used the concept of the socionatural system as a frame of reference in writing about the patterns of land and water use in the Pine Barrens because it is integrative as contrasted to combinatory. Each chapter of *Water, Earth, and Fire* linked people and their uses to local organizations, national markets, state and national legislation, and historical trends, all within the context of local history and ecology. Yet we presented no chapters on marketing systems, the political economy and its relationship to social structure, or changes in the resource base due to economic or political fluctuations. We concentrated instead on environmental ethnography to express both the material and affective aspects of land use. A complete structural analysis of the socionatural system of the region would have obscured this human appreciation of the Pines—a point we wished to impress on practitioners not because it is necessarily more important than any other aspect, but because it is so little appreciated.

Our second perspective requires that planners look at a place from inside as well as from outside, from the affective viewpoint of the users as well as from the scientific view of the academics. Richard Adams, in his 1977 presidential address to the American Anthropological Association, spoke to our concern:

> Why must we continue to be drawn (and indeed sometimes quartered) between the mental and the material? . . . This world incorporates images, sensations, and mentalistic configurations as well as nutritional, chemical, material, and energetic inputs; both are necessary for the survival and well being of man and his society.

Donald Meinig (1970) views this synthesis as "environmental appreciation" or the idea of "localities as humane art." He deplores the overemphasis on the quantitative manipulation of secondary-source data that agglomerates all sense of locality. Roy Rappaport (1968) provides another approach to the same synthesis when he draws the distinction between the cognized and operational models. The cognized model is "the model of the environment conceived by the people who act in it." The operational model is "that which the scientist constructs through observation and measurement of empirical entities, events, and material relationships."

Field Methods

In our fieldwork and research for the Pinelands Commission we used what is best described as naturalistic inquiry (Denzin 1971). It is an umbrella approach that pulls in any technique useful in gathering and understanding data. The method is flexible, technically pragmatic, self-developing, and capable of providing feedback during the course of investigation. In addition, researchers can rely on their own data subjects to test and validate their findings.

We found a combination of methods most useful in assessing local social organization and use patterns. These included an understanding of regional ecological, historical, economic, and cultural contexts, various site-specific natural-resource inventories, an analysis of local published materials, an intensive reconnaissance by car and aerial photography, a directed set of key informant interviews, and participant observation including day-long field trips clamming, hunting, trapping, crabbing, and logging. For four months in the summer and fall of 1980 we employed sixteen fieldworkers and rechecked our fieldwork. At various stages we interpreted and synthesized our data. This mixed-assessment process is called "triangulation" (Denzin 1971) (from the concept in navigation of using several points of reference to determine one's line of travel). Each method has its own data sources, scale of investigation, synthesis procedures, and predictive power. The methods are designed to move down in scale from method to method and to integrate information from the regional to the local level and back to the regional.

Of all these methods we found that planners were least inclined to rely on ethnography or participant observation, which is unfortunate because this cluster of techniques yields crucial information. One of the many excellent summaries of methods is found in Nora Rubinstein's recent work on the Pine Barrens (1983). Pointing to a series of classic studies, such as Whyte's *Streetcorner Society* (1955), Gans's *Urban Villagers* (1962), Liebow's *Talley's Corner* (1967), Coles's *Children of Crisis* (1967), Erikson's *Everything in Its Path* (1976), and Rowles's *Prisoners of Space?* (1978), Rubinstein (1983, 65) finds that

> it is odd that these techniques remain in the methodological "closet" in other [than anthropological] fields of scientific research. But [empathetic] reactions such as Erikson's to the research setting have troubled the scientific community, and the discomfort has resulted in field research, and specifically participant observation being the "step-child" of the methodological repertoire. With the growing body of participant observation literature in fields outside of anthropology, however, it is time that these techniques were recognized as legitimate and valid components of that . . . repertoire, and the content of the research as a valid contribution, not only as exploratory or pilot studies. . . . While it would be convenient, and would probably facilitate the recognition of the participant observation approach as valid, to be able to codify participant observation techniques in a rigorous fashion, the idiosyncratic nature of the field setting must determine the nature of the specific method to be used, and the parameters of "adequate data." This is not to say that there are no accepted techniques and rules for engaging in field research. There is now a literature sufficient to recognize certain techniques as standard. The field method, however, will continue to rely on the creativity of the researcher in the specific setting for much of its information.

Recent reviews that link field methods to other research methods are in Agar (1980), Rubinstein (1983), Burgess (1982), and Spradley (1979; Spradley and McCurdy, 1972). A mix of formal and informal methods should become part of every planner's methodological repertoire. The combined approach is one of the most important ways to link the outsiders' and scientists' perspectives of macro trends to the specific locality for which one is planning. Participant observation also brings to planners a depth of understanding otherwise unobtainable.

A Brief Review and Critique of Environmental Planning

Since the early 1970s, natural-resource planners and other analysts have used a partial and skewed data set to inventory and evaluate human use and organization of the environment. The best plans for determining management practices and future uses usually analyze only a portion of the physical, technical, and cultural system, and pay little attention to the web of interconnections. The method inventories ecological relationships derived from reports, maps, cross sections, fieldwork, and computer models, and the overwhelming proportion of technical reports deals with the physical and biological sciences plus the technological and economic impact of human development. The human aspect of the man / nature system is separated from the nonhuman and depicted by visual analysis, census data, economic statistics, resource mapping, and inventories of historic and prehistoric sites. Through standard planning processes planners add analyses of institutions, legal frameworks, and public participation, and demand forecasts generally based on demographics. These are all desirable and necessary data and processes, yet they fail to deal explicitly with a systemic view of region or to provide an integration of humans and their landscapes.

A review of some useful technical manuals (Clark 1974; Fabos 1975; Fitzpatrick 1978; Hackett 1979; Lewis 1968; Lovejoy 1973; W. Marsh 1978; McEvoy and Dietz 1977; Meshenberg 1976; National Academy of Sciences 1980; Ouellette et al. 1978; Park 1980; Simonds 1978) suggests that, while many environmental planners believe the profession has a theoretical basis, actual practice relies on extensive use of data on the natural environment and inadequate inventories and integration of sociocultural information. Many professionals, especially in the field of environmental-impact assessment, already understand this problem. Witness the comments of Anne Giesecke (1981), an archaeologist with the Water Resources Division of the National Park Service:

> Three important elements to understanding the human environment in a project are: (1) the natural environment; (2) the social or human environment; (3) the history and remains of the material culture or cultural resources in the area. Currently when an environmental impact statement or planning document is being prepared, each of these elements is separated out. Specialists are hired . . . walk over the same ground, address values that are common to all, but never meet physically, mentally or in print. The result of their research is a discontinuous string of technical reports. This information should be provided to decision makers in a useable, integrated form. They should not be expected to piece together sentences from four or five different environmental impact statements in order to see the cumulative effects to the humans and resources of the area.

Such criticisms abound in the literature on impact assessment, and readers need refer to only one of the major journals, such as the *Environmental Impact Assessment Review* or the *Journal of Environmental Management*. Even with extensive technical reports on history and archaeology, decision makers are left with volumes of discrete material compiled by experts who are asked to behave like the proverbial blind men studying the elephant. Planners or consultants rarely look for systematic spatial concurrences or linking processes between landscape and social phenomena; instead, they look for answers, which must be given within a specific, usually inadequate time frame, to questions that may have little bearing on the most difficult planning problems. The best analyses chart, through cross sections, the spatial concurrence of cultural and natural factors, but there is little attention to the range of values and concerns that land users have for specific subregions or patches in the regional mosaic. (See, in particular, the East Everglades Plan, Juneja 1980). Further, without investigation of the relationships among economic, cultural, ecological, historic, and social phenomena, the planning documents may fail to provide usable information needed by participants in the planning process.

The outlook for better theoretical frameworks in planning, however, is not at all bleak, as one can see from a series of attempts to deal with integration more adequately and to critique the present structure over the past decade (Bennett 1980a, 1980b; Dansereau 1975; Hills 1974; Lee 1982; McHarg 1973; Odum 1971; Zube 1982). Let us take a more careful look at some of these contributions.

Zube and his colleagues (1982), in a recent review of landscape-perception research and its application and theory, identified expert, psychophysical, cognitive, and experiential research paradigms (table 6, fig. 11). The expert paradigm involves the evaluation of visual quality by experts trained in art, design, ecology, or resource-management fields where "wise use resource management techniques may be assumed to have intrinsic aesthetic effects." A psychophysical assessment involves the testing of the evaluations of the aesthetic qualities of landscape by the general public or selected populations. A cognitive study of visual quality involves the search for human meaning associated with landscape properties or whole landscapes. Variations occur according to past experience, future expectation, and the cultural milieu of the observer. The experiential paradigm considers landscape value based on the interaction between the human experience and the landscape

TABLE 6. *Model Elements and Landscape-perception Paradigms*

	Expert	Psychophysical	Cognitive	Experiential
Human model	Elite, highly-skilled trained observer	Observer as respondent	Observer as processor	Active participant
Landscape properties	From principles of art, design, ecology, and resource management: Form Balance Contrast Character Diversity Ecological principles — diversity Silviculture — timber stand improvement Pollution control	Specific landscape properties manipulatable through management and design: Cover Water Topography Structures	Associated with obtaining information and meaning: Mystery Legibility Identifiability Prospect Refuge Hazard	World of everyday experience: Familiarity Social space Landscape style
Interaction outcomes	Statement of landscape quality Enhanced sense of landscape	Numerical or statistical expression of perceived values Related landscapes or landscape features	Meaning Ratings of satisfaction — dissatisfaction and preference Stress reduction Adaptation Arousal	Habitual behavior Understanding of human and landscape development Change Statements of landscape taste Enhancement of sense-of-self

SOURCE: *Zube et al. (1982).*

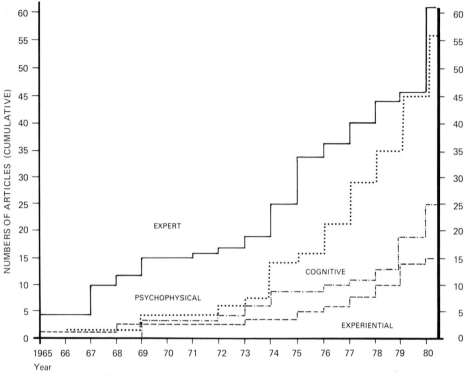

FIGURE 11. *Overall Paradigm Trends*
(reprinted, by permission, from Zube et al., 1982)

whereby the evaluator is an active participant looking at the world of daily experience. Table 6 and figure 11 show the increasing understanding of interactions between users and the landscape as the methods shift toward the experiential paradigm. The cognitive and experiential methods deal explicitly with values and behavior. They can be linked to historical landscape studies, ecosystem models, observation of use patterns, and the various users found on any landscape.

Cognitive and experiential methods, of course, include participant observation; they are little used and even less written about in planning, although they are the most holistic. To use these methods a planner must understand the ecosystem as well as the symbolic and economic values users place on the landscape. The use of the expert paradigm highlights the criticisms and need for a visual analysis integrated with a study of the regional society and its use of the environment. Expert analysis assumes that everyone views the landscape according to the same categories as the expert (e.g., high, medium, or low value). It does not account for regional biases, passing fads, or local variations from national tastes. Expert analysis assumes a consensus on the aesthetic quality of the environment in which all users become "everyone."

Similarly, the mapping of resources—sand and gravel, shellfish, cedar bogs, farmlands, forests, crossroads, etc.—although quite detailed (Fabos 1973; Lewis 1968), provides only a partial picture of the socionatural system. One is left to wonder who uses these places; what are the problems of control, access, and opportunity costs; and what is the relationship between the resource and the well-being of the local community. There is a strange quality about these inventories—a nonhuman quality, for the people who make resources out of these materials and processes are missing. The assumption is that the identification of a resource is the identification of a public value. The resources seem to pop into view without any historical, social, or economic context. Such a picture is obviously incomplete. As Bennett (1976) has suggested, a more complete analysis would include a definition of the physical and technical system into which change is proposed, the amount of balance between the parts, and its capacity for change and readjustment; the action strategies employed by the users and the ways in which they vary by class; the cultural precedents governing the choice of strategy; the nature of innovations or inventions in the system and the way they vary by social and political power; and the relation of the system to wider systems.

The analysis of legal power and institutional responsibility is quite explicit and adequately covered in many studies, such as the National Agricultural Lands Study (Coughlin and Keene 1981), and the East Everglades Study (Juneja 1980). These provide a series of alternatives for environmental protection and land-use regulation. They stop short, however, of evaluating the possible success or failure of any proposed measure. While the Sanibel study may be as good as can be produced at this point, and it had the advantage of the clean slate most new, discrete developments start with, we still need a more complete model in order to evaluate the impact of the planning process. The usual response to this criticism is that such work was not in the contract or that too many unanticipated local factors would have thwarted adequate evaluation. What this may mean is that the planning study does not supply usable information. There is no anticipatory research on the impact of any proposed tool. For example, are all agricultural land-protection strategies equally suited to an agricultural region under urban pressure? In addition to legal and institutional data, analysis of landownership and land-sales patterns, social, economic, and political organization of the regional farming community, patterns of use (soil, water, climate, dispersement of holdings), and family-succession patterns would provide a basis for evaluation.

Again, the use of information on adaptive systems would greatly help anticipate the positive or negative impacts of planning tools. Early and frequent consultation and participation with the human parts of systems (called ''the public'' by professionals) are fundamental in understanding a place and implementing a plan. Public participation (now generally known as public involvement) is one of those hoary platitudes professionals spout, knowing it is necessary if federal and state guidelines are to be satisfied but too difficult and time-consuming to take seriously.

Although literature on public involvement is voluminous, methods to determine or identify publics are themselves sketchy and not conducive to building a holistic regional data base. Creighton (1981) provides readers with a good standard list of methods to identify prospective participants.

All these techniques are useful in concert with other data, yet they relate to the distribution of groups and individuals outside a systematic context. Any inventory of publics based on systems analysis would show links to the environment as well as to the control structure. This needs to be done before the public-involvement program begins, in order to give the planner an understanding of the systematic workings of the place and later to help him evaluate the appropriate tools and the relative success of the plan.

Rather than the piecemeal approach embedded in the current treatments of the ''human environment,'' we advocate that a portion of the planning report cover a section on environmental values or user/environment relationships. Such a section would cover historical changes in the socionatural system as seen through the history of settlement, resource use, migration, demographics, and economic structure. The historical treatment would document the cultural precedents for current patterns of use and for the changing economic and institutional base; it would also provide a framework in which to see the relationships of small (micro) localities to regional and national or international (macro) systems. Data on contemporary use should include information on seasonality, periodicity, range and distribution of resource uses, methods of extraction, and the political, economic, and cultural importance of the resource in both an ecological and a cultural context.

A combination of historical and ethnographic information provides the background for under-standing: traditional techniques of use and management; local and subregional aesthetic norms; local and regional attitudes toward political, social, and economic control of resources; and the content and resolution of conflicts over resource use. Our proposed synthesis would provide an understanding of long traditions of use and belief and their relationship to the environment and to the quality of life—the social and mental health of local communities. Each land use or complementary cluster of land uses would then be understood in terms of exploitive technology, development activities, and the social organization or political economy of the place as well as the impacts of those land-use patterns on the social and natural environment, the full range of participants, and the symbolic and aesthetic meanings of those uses. There would be no separation of humans and nature so common to most resource-management plans. The separation continues to this day in the Pinelands planning process, as is indicated by the proceedings of a 1982 conference on "Ecological Solutions to Environmental Management Concerns in the Pinelands National Reserve" sponsored by the National Science Foundation, Office of Water Resources Technology. Of the thirty-five participants, there were no anthropologists, economists, historians, or political scientists, yet the goal was to provide "an implementation guide for managers of the Pinelands" (Good 1982, vii).

Although our suggestions for synthesis do not occur in the data base for most environmental plans or in the recommendations of a host of technical manuals, seminal thinkers, such as Benton MacKaye, have either practiced or endorsed such concepts. Sir Patrick Geddes believed that planning and design entailed "Sympathy, Synthesis and Synergy"; he recommended sympathy for the people and their landscapes who were affected by social remedies, such as planning, and asked for a synthesis of all factors relevant to a planning situation. Geddes, believing that synergy involved the combined cooperative action of everyone in order to achieve the best result, called for an understanding of the history, folklore, and community sense "inlaid in the old fabric." He practiced his belief that the collection of data preceding a planning or design exercise was

> a means toward the realization of our community's life history. The life history is not past and close with it is incorporated all present activities and characteristics. . . . From our survey of facts we have to prepare no mere material record, economic or structural, but to evoke the social personality, changing indeed so far with every generation yet ever expressing itself in and through these. (Geddes in Stalley 1972)

Geddes was a fieldworker and a participant in the cities he helped reconstruct. He was a professor of biology and sociology and understood the earth as well as he did human uses and people's dreams. He was dedicated to creative change and called planning "geotechnics" ("geo" for the life system of the earth, and "technics" for the restoration, design, and understanding of that system).

The late Canadian planner Angus Hills, also a prodigious fieldworker and landscape interpreter, produced a holistic philosophy and method of ecological planning. In one of his last essays before his death, Hills (1974) wrote:

> It is geographic totality—not man, nor nature, nor physiography, which provides the nexus for planning. Within this pattern of landscapes, possible alternatives are established by examining various matching combinations of physiographic, natural, and social determinants.

Only historical accident—the spectacular rise of public concern about people's deleterious impacts on the earth's ecology—diverted the generation of planners and teachers of the sixties and seventies from the larger investigation of socionatural systems and from the incorporation of a broad humanistic appreciation of landscapes. We suggest that landscape or ecological planners not only expand their model, but return to the initial directions set by our mentors.

Toward a New Environmental-planning Framework

Our approach raises important theoretical questions about the relationship between the sociocultural environment, the natural and built environments, the mental health of individuals and communities, and the translation of such a theory and its accompanying data into planning language.

Of all these questions, the most critical is what John Bennett termed "the micro-macro problem" in a seminal essay titled "The Micro-Macro Nexus" (1983). The "problem" is clearly not limited to anthropologists; one of Bennett's assumptions is "the necessity for anthropology to join with other social disciplines in dealing with the demanding issues of world social transition" (p. 1). The authors would have to include the natural sciences and applied disciplines, such as planning, in searching for solutions. Bennett defined the problem as follows:

> I have found it useful to condense the many available topics into just three general ones: (1) "micro" and "macro" phenomena defined as different types of community, society, or culture; (2) "micro" and "macro" as referring to interactive processes of change, involving diffusion, influence, and defensive adaptation; and (3) the question of broader systemic convergence in world society: how "micro" social phenomena, and/or local communities, are being incorporated into ever-larger systems of action and control. . . . Another comment: my choice of the word "nexus" refers to a major emphasis on process. I consider that the main task is to determine how the micro (or local) forms interact with the macro (or external) forms: how people solve their problems in a milieu of relationships and adaptive coping. I further consider that this problem of nexus is becoming the future theme of sociocultural anthropological research; it is the mainline anthropology of the future. And once attention turns to such issues, the ethical and ideological relevance of our work emerges in concrete and unavoidable form. . . . The essential feature of a theoretical paradigm is the recognition of the need for considering the will, purposes, and needs of local-level people in the bureaucratic institutions which govern and regulate the world. . . . The second theme concerns moral obligations to humanity. . . . Whatever [the anthropologist] does, it has inevitable practical and policy relevance, because the people he studies are also *in* the world. Thus the moral obligation is not a matter of choice; it is necessity, thrust on the anthropologist by his participation in a changing society. . . . This is the final significance of the micro-macro frame: it focusses our attention on what is really going on in the world: the incorporation of communities into larger and larger systems, a process which needs research at both the micro level, to determine what it is doing to people and their groups, and also at the macro level, to find out how these enormous human constructions are being put together. We know little about either end of the continuum, and even less about the nexi that tie the ends together. (1–3, 51–55)

In dealing with complex systems, whether national reserves or international structures, the keys to their understanding lie in how the levels are connected to each other, be they individuals of a biological species in its larger ecosystem or a human being within his or her national, even international, context. Jane Jacobs has recently made a major contribution with *Cities and the Wealth of Nations* (1984).

Let us look more closely, albeit briefly, at what this means in a planning context. We have suggested in the body of *Water, Earth, and Fire* the extent to which planners get wrapped up in "issues" and conflict resolution, and how concentration on such issues forces planners to deal in the immediate present, allowing them to disregard the importance of larger frameworks. A number of important planners, however, have tried to develop larger contexts; we have already mentioned Hills, Geddes, McHarg, and MacKaye, and there are others, not the least of whom are Frederick Law Olmsted, Lewis Mumford, and Jane Jacobs. However, the single context in which planners generally deal with micro-macro problems is political, because planning must take place in political arenas and the political process is the one most often used in conflict resolution in Western democracies. Unfortunately, such strong reliance on political procedures obviates the use of other methods to solve or avoid conflicts, and, more important, so simplifies a planner's vision of a region that it distorts the real complexity of a complex system. In short, what we are saying is that a regional plan's tendency is to limit, rather than expand, the courses of action open to people and their landscapes.

An aspect of this problem has been graphically described in John Forester's work (1980, 1982) which deals with communication, misinformation, and power. Forester (1980) suggests that a technically oriented planner, no matter how well intentioned, can unwittingly subvert his or her own efforts:

Ironically . . . technically-oriented planning may effectively but unintentionally communicate to the public, "you can depend on me; you needn't get involved; I'll consult you when appropriate." This message may simplify practice in the short-run, but it may also lead to inefficiency and waste in general. It counterproductively may separate planners from the political constituency they serve, weakening them both before the designs and agendas of powerful economic forces in their neighborhoods and cities. It may subvert the accountability of planners and serve to keep affected publics uninformed rather than politically educated about events and local decisions affecting their lives. (282)

What we need, and do not yet have, is a framework in which to understand the workings of a region from its highest to lowest level and the ways in which these levels interact. Meanwhile, we must continue to work on all levels to piece together visions of the way places work.

Finally, we would like to suggest those levels on which planners need to work hardest. In planning we know quite a lot about what occurs on the micro level in the natural sciences (although groundwater hydrology presents problems and will continue to do so if for no other reason than its location underground). Similarly, ecological knowledge is quite sophisticated, although we do not know a great deal about the relationship between the size of ecosystems and their species diversity or the relative size requirements for each species—again a micro-macro problem. But if one has followed environmental planning over the past two decades, with its development of sophisticated models, one cannot help but wonder why in comparison so little is known of human systems. The natural sciences, in any case, tend to be ahead of other disciplines in terms of their understanding of relationships among different levels of energy or organization.

With this and similar studies of socionatural systems, we hope that planners will begin to bridge that gap between humans and their landscapes. We have had much trouble in understanding the relationships of individuals to a variety of contexts, and this goes beyond public-involvement techniques. In general, the planners' grasp of the human environment is rudimentary, and, as we stated before, based almost solely on operational models. Such models are convenient because they allow the expert to piece together sets of socioeconomic, political, and legal data, thus merging the individual into an interest group to be dealt with in a political context.

Up to the present, environmental psychologists have given planners little help (Wohlwill 1978). The micro level of the individual, critical though it is, has continued to elude us. Individuals are not only ambivalent about their lives and landscapes, but, as Silvan Tomkins has stated, they are plurivalent—of many minds (personal communication, March 1983). Plurivalence is anathema to planners because it stymies quick decisions and obscures the neatness of planning solutions. On the other hand, it often does not help to place individuals into convenient categories, such as environmentalists, farmers, developers, or Pineys. Rubinstein (1983) described the problem vividly:

There are many groups which lobby for the environmental "preservation" of the Pines. A short list of the groups which have testified before the Commission includes: The Pine Barrens Coalition, the Sierra Club (both local and national chapters), the Association of New Jersey Environmental Commissions, the Natural Resources Defense Council, the New Jersey Conservation Foundation, the Environmental Defense Fund, the Wild and Scenic Rivers Coalition, and the Audubon Society. It is clear from past performance that not only do the groups have different positions on specific issues, but it can be difficult to achieve consensus on the over-riding issue of the definition of "protection." Certainly individual members don't ascribe to all the positions of all "environmental lobby groups."

Neither do all developers want to pave over the Pines as is claimed by some of the environmental activists. . . . Farmers are not monolithic either. . . . Neither is the label Piney uni-dimensional. There are almost as many different types of Pineys as there are people to define them.

If preservation is an ambiguous term in the ecological context, it is no less so in social terms. A legislative mandate to protect or preserve while permitting "appropriate growth" is a semantic disaster. The nature of protection is undefined, as is appropriate growth, and this dilemma has been at the base of the conflicts between farmers, developers and environmentalists ever since. (38–39, 55)

It is true that planning cannot respond solely to the confusing evidence one finds in the minds of individual residents, but to deny such evidence is unethical and self-defeating. Our problem, then, is to find methods to translate data from individuals into the planning context. Nora Rubinstein's dissertation (1983) studied the psychosocial impacts of change in the Pine Barrens. Her work includes valuable sections on methodology, values, and identity and defines the implications of environmental psychology for planning. Further cooperation among planners and psychologists will develop a significant body of similar studies and advance our ability to understand relationships on a micro level and develop better techniques for implementation.

Other levels of sociocultural and economic analysis need to be strengthened as well, although planners have better access to the tools of regional economic or social analysis than to the work of psychologists. We should have more studies such as those of the University of Maine anthropologist James Acheson, who has been studying the anthropology of fishing (1981), and Phillip Miller of the University of Toronto, who created a simulation model for a northwest coastal fishing village (1982, 143). Miller concluded with a question: "Are there then, any properties of complex systems that underlie development in human settlements?"

We have tools to enhance our comprehension of micro-macro problems; they are simply underutilized. Less known than these middle levels of reality are the workings of the largest systems. We need more work on how to analyze local impacts from national and international perturbations, and, of course, the ways in which these levels are interconnected.

Above all, we need a new framework in which we can view our work and through which scholars and practitioners in all disciplines can communicate. We will arrive at that point, but we will have to change our way of looking at the world somewhat; in fact, we will have to become more accustomed to accepting the simple fact of change.

References

Acheson, J. M. 1981. Anthropology of Fishing. *Annual Review of Anthropology* 10:275–316.

Acheson, J. M., and R. Reidman. 1982. Biological and Economic Effects of Increasing the Minimum Legal Size of American Lobster in Maine. *Transactions of the American Fisheries Society* 111, no. 1:1–12.

Adams, R. N. 1978. Man, Energy, and Anthropology: I Can Feel the Heat, But Where's the Light? *American Anthropologist* 80, no. 2:297–309.

Agar, M. 1980. *The Professional Stranger: An Informal Introduction to Ethnography.* New York: Academic Press.

American Rivers Conservation Council. 1981. Summer Calendar 1981. *American Rivers Conservation Council Report.* Washington, D.C.

Applegate, J. E., S. Little, and P. Marucci. 1979. Plant and Animal Products of the Pine Barrens. In Forman 1979, 25–36.

Ayres, T. 1979. The Pinelands Cultural Society: Folk Music Performance and the Rhetoric of Regional Pride. In Sinton 1979, 225–60.

Bateson, G. 1972. *Steps to an Ecology of Mind.* San Francisco: Chandler.

Bennett, J. 1976. *The Ecological Transition: Cultural Anthropology and Human Adaptation.* London: Pergamon.

———. 1980a. Human Ecology as Human Behavior: A Normative Anthropology of Resource Use and Abuse. In *Human Behavior and Environment.* Vol. 4, *Environment and Culture,* edited by I. Altman et al. New York: Plenum.

———. 1980b. Social and Interdisciplinary Sciences in U.S. *MAB* [Man and the Biosphere Program]: Conceptual and Theoretical Aspects. In *Social Sciences, Interdisciplinary Research, and the U.S. Man and the Biosphere Program: Workshop Proceedings,* edited by Ervin Zube.

———. 1983. The Micro-Macro Nexus: Typology, Process, and System. Presented at the 1981 Annual Meeting of the American Anthropological Association. To be published in a book edited by P. J. Pelto and B. De Walt.

Berger, J. 1980. *Planning the Use and Management of the Pinelands: A Cultural, Historical, and Ecological Perspective.* New Lisbon, N.J.: Pinelands Commission, unpublished report.

Berger, T. R. 1977. *Northern Frontier, Northern Homeland: The Report of the Mackenzie Valley Pipeline Inquiry,* vol. 1. Ottawa: Minister of Supply and Services.

Bucholz, K., and R. Good. 1982. *Survey of Current Research on the New Jersey Pine Barrens.* New Brunswick: Rutgers University Center for Coastal and Environmental Studies.

Burgess, R. G., ed. 1982. *Field Research: A Sourcebook and Field Manual.* London: Allen & Unwin.

Cawley, J., and M. M. Cawley. 1961. *Exploring the Little Rivers of New Jersey.* New Brunswick: Rutgers University Press.

Clark, J. 1974. *A Technical Manual for the Conservation of the Coastal Zone.* New York: Wiley.

Coggins, C. G. 1980. Wildlife and the Constitution: The Walls Come Tumbling Down. *Washington Law Review* 55:295.

Coles, R. 1967. *Children of Crisis.* Boston: Little, Brown.

Coughlin, R. E., and J. C. Keene. 1981. *The Protection of Farmlands: A Reference Guidebook for State and Local Governments.* Washington, D.C.: U.S. Government Printing Office no. 0–335–616.

Creighton, J. L. 1981. *The Public Involvement Manual.* Cambridge, Mass.: Abt Books.

Crerar, D., G. Knox, and J. L. Means. 1979. Biogeochemistry of Bog Iron in the New Jersey Pine Barrens. *Chemical Geology* 24, no. 1/2:111–15.

Cross, D. 1965. *New Jersey's Indians.* Trenton: N.J. State Museum.

Cunz, D. 1955? *Egg Harbor City: New Germany in New Jersey.* Atlantic City: n.p.

Dansereau, P. 1975. *Harmony and Disorder in the Canadian Environment.* Ottawa: Canadian Environmental Advisory Council, Occasional Paper no. 1.

Denzin, N. K. 1970. *The Research Act.* Chicago: Aldine.

———. 1971. The Logic of Naturalistic Inquiry. *Social Forces* 2:169–81.

Erikson, K. 1976. *Everything in Its Path.* New York: Simon & Schuster.

Fabos, J. (Gy.) 1975. *Planning the Total Landscape: A Guide to Intelligent Land Use.* Boulder, Colo.: Westview Press.

Figley, W., and T. McCloy. 1980. *New Jersey's 1980 Bay Shellfisheries.* Trenton, N.J.: Division of Fish, Game, and Shellfisheries.

Fitzpatrick, M. S. 1978. *Environmental Health Planning.* Boston: Ballenger.

Forester, J. 1980. Critical Theory and Planning Practice. *Journal of the American Planning Association* 46, no. 3:275–86.

———. 1982. Planning in the Face of Power. *Journal of the American Planning Association* 48, no. 1:67–80.

Forman, R. T. T., ed. 1979. *Pine Barrens: Ecosystem and Landscape.* New York: Academic Press.

Forman, R. T. T., and R. E. Boerner. 1981. Fire Frequency and the Pine Barrens of New Jersey. *Bulletin of the Torrey Botanical Club* 108, no. 1:34–50.

Gans, H. 1962. *Urban Villagers.* New York: Free Press.

Giesecke, A. 1981. The Human Environment Has a Context. *Environmental Impact Assessment Review* 2, no. 2:137–39.

Gillespie, A. 1980. Foodways in the Pine Barrens of New Jersey. New Brunswick: Douglass College. Unpublished manuscript.

Goddard, H. H. 1912. *The Kallikak Family.* New York: Macmillan.

Goldstein, J. 1981. *Environmental Decision Making in Rural Locales: The Pine Barrens*. New York: Praeger.

Good, R., ed. 1982. *Ecological Solutions to Environmental Management Concerns in the Pinelands National Reserve: Proceedings of a Conference*. New Brunswick: Rutgers University Center for Coastal and Environmental Studies, Division of Pinelands Research.

Good, R., N. Good, and J. Andresen. 1979. The Pine Barren Plains. In Forman 1979, 283–95.

Gordon, T. 1982. The Last of the Old-time Charcoal Makers and the Coaling Process in the Pine Barrens of New Jersey. In Sinton 1982, 212–21.

Gordon, T. 1834. *A Gazeteer of the State of New Jersey*. Trenton: D. Fenton.

Gould, S. J. 1981. *The Mismeasure of Man*. New York: Norton.

Hackett, B. 1971. *Landscape Planning: An Introduction to Theory and Practice*. Newcastle upon Tyne: Oriel.

Halpert, H. 1947. Folk Tales and Legends from the New Jersey Pine Barrens: A Collection and a Study. Ph.D. dissertation, Indiana University.

Harshberger, J. 1916. *The Vegetation of the New Jersey Pine Barrens, an Ecologic Investigation*. Philadelphia: C. Sower.

Hartzog, S. 1982. Palynology and Late Pleistocene-Halocene Environment on the N.J. Coastal Plain. In Sinton 1982, 6–14.

Hills, A. 1974. A Philosophical Approach to Landscape Planning. *Landscape Planning* 1, no. 4:339–71.

Hufford, M. 1982. Foxhunting in the Pine Barrens. In Sinton 1982, 222–34.

Juneja, N. 1980. *East Everglades Resources Planning Project*. Miami: Dade County Planning Department and Wallace, Roberts & Todd.

Kauffeld, C. F. 1957. *Snakes and Snake Hunting*. Garden City, N.Y.: Hanover House.

Koh, J. 1982. Ecological Design: A Post-modern Design Paradigm of Holistic Philosophy and Evolutionary Ethic. *Landscape Journal* 1, no. 2:76–84.

Lee, B. J. 1982. An Ecological Comparison of the McHarg Method and other Planning Initiatives in the Great Lakes Basin. *Landscape Planning* 9:147–69.

Leopold, A. 1949. *A Sand County Almanac*. New York: Oxford University Press.

Lewis, P. 1968. *Upper Mississippi River Comprehensive Basin Study*. Appendix B: Aesthetic and Cultural Values. Prepared under the auspices of the U.S. Dept. of the Interior, National Park Service, Northeastern Region, UMRB Coordinating Committee, Kaukauna, Wis.

Liebow, E. 1967. *Talley's Corner*. Boston: Little, Brown.

Little, S. 1979a. Ecology and Silviculture of Pine Barrens Forests. In Sinton 1979, 105–18.

———. 1979b. Fire and Plant Succession in the New Jersey Pine Barrens. In Forman 1979, 297–314.

Lovejoy, D. 1973. *Land Use and Landscape Planning*. New York: Barnes & Noble.

MacKaye, B. 1928. *The New Exploration*. New York: Harcourt, Brace.

Mallach, A. 1980a. *Growth Shapers in the Pinelands*. New Lisbon, N.J.: Pinelands Commission, unpublished report.

———. 1980b. *Land Market and Land Development Trends in the Pinelands*. New Lisbon, N.J.: Pinelands Commission, unpublished report.

Marsh, E. 1979. The Southern Pine Barrens: An Ethnic Archipelago. In Sinton 1979, 192–98.

———. 1982. The South Jersey House. In Sinton 1982, 185–92.

Marsh, W. 1978. *Environmental Analysis for Land Use and Site Planning*. New York: McGraw-Hill.

McCormick, J. 1970. *The Pine Barrens: A Preliminary Ecological Inventory*. Trenton: New Jersey State Museum.

———. 1979. The Vegetation of the New Jersey Pine Barrens. In Forman 1979, 229–43.

McEvoy, K., and L. Dietz. 1977. *Handbook for Environmental Planning: The Social Consequences of Environmental Change*. New York: Wiley.

McHarg, I. 1981. Human Ecological Planning at Pennsylvania. *Landscape Planning* 18, no. 2:109–20.

_____. 1969. *Design with Nature*. Garden City: Natural History Press.

McPhee, J. 1968. *The Pine Barrens*. New York: Farrar, Strauss and Giroux.

Meinig, D. R. 1970. Environmental Appreciation: Localities as Humane Art. *Western Humanities Review* 25:1–11.

_____. 1972. American Wests: Preface to a Geographical Interpretation. *AAAG*, no. 62:159–84.

Meshenberg, M. J. 1976. *Environmental Planning: A Guide to Information Sources*. Detroit: Gale.

Miller, P. C. 1982. Simulation of Socio-ecological Impacts: Modeling a Fishing Village. *Environmental Management* 6, no. 2:123–44.

Mounier, R. A. 1982. The Late Woodland Period in Southern New Jersey. In Sinton 1982, 116–38.

National Academy of Sciences, Environmental Studies Board, Committee on Urban Waterfront Lands. 1980. *Urban Waterfront Lands*. Washington, D.C. National Research Council.

New Jersey Department of Environmental Protection, Division of Fish, Game, and Shellfisheries. 1978. *Upland Wildlife and Habitat Investigations, no. XVI—Harvest, Recreational, and Economic Surveys* (Project no. W–52–R–6). Trenton.

_____, Bureau of Forestry. 1980. *Directory of New Jersey Primary Wood Processors and Users*. Trenton.

New Jersey Senate and General Assembly. 1979. P.L. 1979, chap. 111 (The Pinelands Protection Act).

New Jersey State Legislature. 1886. *Acts of the American Legislature*. Trenton.

Odum, H. T. 1971. *Environment, Power, and Society*. New York: Wiley.

Oulette, S. et al. 1978. *Environmental Impact Data Book*. Ann Arbor, Mich.: Ann Arbor Science.

Park, C. 1980. *Ecology and Environmental Management*. Boulder, Colo.: Westview Press.

Philadelphia, city of, Commission of Engineers. 1875. *Report on the Water Supply of the City of Philadelphia*.

Pierce, A. 1957. *Iron in the Pines*. New Brunswick, N.J.: Rutgers University Press.

_____. 1964. *Family Empire in Jersey Iron*. New Brunswick, N.J.: Rutgers University Press.

Pierson, G. 1979. The Wood-Using Industries of the Pinelands. In Sinton 1979, 119–31.

Pinelands Commission. 1980. *New Jersey Pinelands Comprehensive Management Plan*. New Lisbon, N.J.: Pinelands Commission.

Rappaport, Roy. 1968. *Pigs for the Ancestors*. New Haven, Conn.: Yale University Press.

Rees, R. 1975. The Scenery Cult: Changing Landscape Tastes over Three Centuries. *Landscape* 19:39–47.

Regensburg, R. 1979. Evidence of Indian Settlement Patterns in the Pine Barrens. In Sinton 1979, 199–213.

Rhodehamel, E. C. 1979. *A Hydrologic Analysis of the New Jersey Pine Barrens Region*. New Jersey Water Resources Circular 22. Trenton.

Robichaud-Collins, B., and E. Russell. n.d. *Preserving the Pinelands of New Jersey*. In preparation.

Robinson, A. 1980. *Surface Water Quality*. New Lisbon, N.J.: Pinelands Commission (Technical memorandum to report by Betz, Converse, and Murdoch, Inc.).

Rose, D. 1979. The Aesthetic and Moral Ordering of the Material World in Southern Chester County, Pennsylvania. *Anthropology and Humanism Quarterly* 4:14–21.

Rose, T. F., and H. C. Woolman. 1878. *Historical and Biographical Atlas of the New Jersey Coast*. Philadelphia.

Ross, Hardies, O'Keefe, Babcock, and Parsons. 1980. *Procedural and Substantive Land Management Techniques of Potential Relevance for the New Jersey Pinelands*. 5 vols. New Lisbon, N.J.: N.J. Pinelands Commission.

Rowles, G. 1978. *Prisoners of Space? Exploring the Geographical Experience of Older People*. Boulder, Colo.: Westview Press.

Rubinstein, N. 1983. A Psycho-Social Impact Analysis of Environmental Change in the New Jersey Pine Barrens. Ph.D. dissertation, City University of New York.

Salter, J. F. 1981. Shadow Forks: A Small Community's Relationship to Ecology and Regulation. Ph.D. dissertation, University of California, Santa Cruz.

Simonds, J. D. 1978. *Earthscape: A Manual of Environmental Planning*. New York: McGraw-Hill.

Sinton, J. W. 1977. The Phoenix of the Pines. *Environmental Review* 4, no. 77:17–25.

———, ed. 1979. *Natural and Cultural Resources of the New Jersey Pine Barrens: Inputs and Research Needs for Planning*. Pomona, N.J.: Stockton State College.

———. 1980. Cultural Self-preservation: Planning for Local Cultures in the New Jersey Pine Barrens. *New Jersey Folklore* Spring:12–17.

———, ed. 1982. *History, Culture, and Archeology of the Pine Barrens: Essays from the Third Pine Barrens Conference*. Pomona, N.J.: Stockton State College.

Sinton, J. W., and G. Masino. 1979. A Barren Landscape, a Stable Society: People and Resources of the Pine Barrens in the 19th Century. In *Natural and Cultural Resources of the New Jersey Pine Barrens,* ed. J. W. Sinton. Pomona, N.J.: Stockton State College.

Skinner, A. B., and M. Schrabisch. 1913. *A Preliminary Report of the Archeological Survey of the State of New Jersey*. Trenton: New Jersey State Geological Survey, bulletin 9.

Spradley, J. P. 1979. *The Ethnographic Interview*. New York: Holt, Rinehart & Winston.

Spradley, J. P., and D. W. McCurdy. 1972. The Cultural Experience: Ethnography in Complex Society. Chicago: Research Associates Publishers.

Stalley, M., ed. 1972. *Patrick Geddes, Spokesman for Man and the Environment*. New Brunswick, N.J.: Rutgers University Press.

Stilgoe, J. R. 1981. Fair Fields and Blasted Rock: American Land Classification Systems and Landscape Aesthetics. *American Studies* 22, no. 1:21–33.

Stokes, S. 1980. *Establishing an Easement Program to Protect Scenic, Historic, and Natural Resources*. Washington, D.C.: National Trust for Historic Preservation, Information Sheet #25.

Stone, W. 1911. *The Plants of Southern New Jersey with Especial Reference to the Flora of the Pine Barrens and the Geographic Distribution of the Species*. New Jersey State Museum, Annual Report of the State Geologist.

Thompson, M. A. 1980. Public Use of Private Land in the Pine Barrens. In *Opportunities for Enhancing Public Benefit from Private Land*. Washington, D.C.: American Bar Association, Division of Public Services.

———. 1982. The Landscape of Cranberry Culture. In Sinton 1982, 193–211.

Tillet, P. 1961. *Doe Day*. New Brunswick, N.J.: Rutgers University Press.

Tolles, F. 1948. *Meeting House and Counting House: The Quaker Merchants of Colonial Philadelphia, 1682–1783*. Philadelphia: University of Pennsylvania Press.

U.S. Congress (95th). 1978. Public Law 95–625 (National Parks and Recreation Act of 1978).

U.S. National Marine Fisheries Service, 1977. *New Jersey Landings, Annual Summaries*. Washington, D.C.

Walker, J. 1966. *Hopewell Village: A Social and Economic History of an Iron-making Community*. Philadelphia: University of Pennsylvania Press.

Warner, W. 1976. *Beautiful Swimmers*. Boston: Little, Brown.

Whyte, W. 1955. *Streetcorner Society*. Chicago: University of Chicago Press.

Williams, H., and J. La Rocca. 1979. Some Thoughts on the Pinelands. Unpublished report.

Wohlwill, J. 1978. A Psychologist Looks at Land Use. *Environmental Review* 3, no. 1:34–48.

Wolgast, L. 1977. Effects of Relative Humidity at Time of Flowering on Fruit Set in Bear Oak (*Quercus ilicifolia*). *American Journal of Botany* 64:2.

———. 1978. Effects of Site Quality and Genetics on Bear Oak Mast Production. *American Journal of Botany* 65:4.

Woodford, E. 1970. *A Home in the Pine Barrens*. Medford, N.J.: Medford Township Environmental Commission.

Zube, E. et al. 1982. Landscape Perception: Research Applications and Theory. *Landscape Planning* 9:1–33.

Index

The Johns Hopkins University Press

WATER, EARTH, AND FIRE

This book was composed in Times Roman text and Optima display type by Brushwood Graphics Studio, Baltimore, Maryland, from a design by Cynthia W. Hotvedt.

It was printed on 70-lb. Paloma coated matte paper and bound in Joanna Arrestox cloth by The Maple Press Co., York, Pennsylvania.